AN INTRODUCTION TO GENETICS

AN INTRODUCTION TO GENETICS

Mahipal Singh Rana

CYBER TECH PUBLICATIONS
4264/3, Ansari Road, Daryaganj, New Delhi-110002 (India)
Ph.: 011-23244078. Fax: 011-23280028
E-Mail: cyberpublicationsdelhi@yahoo.com

AN INTRODUCTION TO GENETICS

Mahipal Singh Rana

First Edition 2009

Published by :

G.S. Rawat for **Cyber Tech Publications**
4264/3, Ansari Road, Daryaganj, New Delhi-110002 (India)
Ph.: 011-23244078. Fax: 011-23280028
E-Mail: cyberpublicationsdelhi@yahoo.com

Printed at Asian Offset Printers

Preface

Although the science of genetics has its origins in the work of Gregor Mendel in the mid-nineteenth century, various theories of inheritance preceded Mendel. These theories generally assumed that there existed an inheritance of acquired characteristics: the belief that individuals inherit traits that have been strengthened in their parents. Today, the theory is commonly associated with Jean-Baptiste Lamarck, who used this pattern of inheritance to explain the evolution of various traits within species (these changes are now understood to be the product of natural selection rather than a product of soft inheritance).

The modern science of genetics traces its roots to the observations made by Gregor Johann Mendel, a German-Czech Augustinian monk and scientist who made detailed studies of the nature of inheritance in plants. In his paper "Versuche über Pflanzenhybriden" ("Experiments on Plant Hybridization"), presented in 1865 to the Brunn Natural History Society, Gregor Mendel traced the inheritance patterns of certain traits in pea plants and showed that they could be described mathematically. Although not all features show these patterns of Mendelian inheritance, his work suggested the utility of the application of statistics to the study of inheritance.

—Author

CONTENTS

1

INTRODUCTION

Genetics, a discipline of biology, is the science of heredity and variation in living organisms. Knowledge of the inheritance of characteristics has been implicitly used since prehistoric times for improving crop plants and animals through selective breeding. However, the modern science of genetics, which seeks to understand the mechanisms of inheritance, only began with the work of Gregor Mendel in the mid-nineteenth century. Although he did not know the physical basis for heredity, Mendel observed that inheritance is fundamentally a discrete process where specific traits are inherited in an independent manner—these basic units of inheritance are now called genes.

Genes correspond to regions within DNA, a molecule composed of a chain of four different types of nucleotides—the sequence of these nucleotides is the genetic information organisms inherit. DNA naturally occurs in a double stranded form, with nucleotides on each strand complementary to each other. Each strand can act as a template for synthesis of a new partner strand—this is the physical mechanism for the copying and inheritance of genetic information.

The sequence of nucleotides in DNA is used by cells to produce specific sequences of amino acids, creating proteins—a correspondence known as the genetic code. This sequence of amino acids in a protein determines how it folds into a three-dimensional structure, this structure is, in turn, responsible for the protein's function. Proteins carry out almost all the functions needed for cells to live and reproduce. A change to DNA sequence can change a protein's structure and behavior, and this can have dramatic consequences in the cell and on the organism as a whole.

Although genetics plays a large role in determining the appearance and behavior of organisms, it is the interaction of genetics with the environment an organism experiences that determines the ultimate outcome. For example, while genes play a role in determining a person's height, the nutrition and health that person experiences in childhood also have a large effect.

HISTORY

Although the science of genetics has its origins in the work of Gregor Mendel in the mid-nineteenth century, various theories of inheritance preceded Mendel. These theories generally assumed that there existed an inheritance of acquired characteristics: the belief that individuals inherit traits that have been strengthened in their parents. Today, the theory is commonly associated with Jean-Baptiste Lamarck, who used this pattern of inheritance to explain the evolution of various traits within species (these changes are now understood to be the product of natural selection rather than a product of soft inheritance).

MENDELIAN AND CLASSICAL GENETICS

The modern science of genetics traces its roots to the observations made by Gregor Johann Mendel, a German-Czech Augustinian monk and scientist who made detailed studies of the nature of inheritance in plants. In his paper "Versuche über Pflanzenhybriden" ("Experiments on Plant Hybridization"), presented in 1865 to the Brunn Natural History Society, Gregor Mendel traced the inheritance patterns of certain traits in pea plants and showed that they could be described mathematically. Although not all features show these patterns of Mendelian inheritance, his work suggested the utility of the application of statistics to the study of inheritance.

The significance of Mendel's observations was not understood until early in the twentieth century, after his death, when his research was re-discovered by other scientists working on similar problems. The word *genetics* itself was coined in 1905 by William Bateson, a significant proponent of Mendel's work, in a letter to Adam Sedgwick. (The adjective *genetic*, derived from the Greek word *genno* (*γåίίþ*): to give birth, predates the noun and was first used in a biological sense in 1860.) Bateson publicly promoted and popularized usage of the word *genetics* to describe the study of inheritance in his inaugural address to the Third International Conference on Plant Hybridization in London, England, in 1906.

In the decades following rediscovery and popularization of Mendel's work, numerous experiments sought to elucidate the molecular basis of DNA. In 1910 Thomas Hunt Morgan argued that genes reside on chromosomes, based on observations of a sex-linked white eye mutation in fruit flies. In 1913 his student Alfred Sturtevant used the phenomenon of genetic linkage and the associated recombination rates to demonstrate and map the linear arrangement of genes upon the chromosome.

Molecular genetics

Although chromosomes were known to contain genes and composed of both protein and DNA, it was unknown which was critical for heredity or how the process occurred. In 1928, Frederick Griffith published his discovery of the phenomenon of transformation; sixteen years later, in 1944, Oswald Theodore Avery, Colin McLeod and Maclyn McCarty used this phenomenon to isolate and identify the molecule responsible for transformation as DNA. The Hershey-Chase experiment in 1952 identified DNA (rather than protein) as the genetic material of the viruses that infect bacteria, further evidence that DNA was the molecule responsible for inheritance.

James D. Watson and Francis Crick solved the structure of DNA in 1953, using the X-ray crystallography work of Rosalind Franklin that indicated the molecule had a helical structure. Their double-helix model paired a sequence of nucleotides with complementary nucleotides on the other strand. This structure not only provided a physical explanation for how information is contained within the order of the nucleotides, but also suggested a simple mechanism for duplication—through separation of strands and the reconstruction of partner strands based on the nucleotide sequences. Although the structure explained the process of inheritance, it was still unknown how DNA influenced the behavior of cells. In the following years scientists sought to understand how DNA controls the process of protein production within ribosomes, eventually discovering the transcription of DNA into messenger RNA and cracking the genetic code, which links the nucleotide sequence of messenger RNA to the amino acid sequence of protein.

With this molecular understanding of inheritance, an explosion of research that applied this new knowledge to biology became possible. One early development was chain-termination DNA sequencing in 1977, which enabled the determination of nucleotide sequences on DNA. In 1983 the polymerase chain reaction, developed by Kary Banks Mullis, provided a simple method for isolating and amplifying segments of DNA. These and other techniques, through the pooled efforts of the Human Genome Project and parallel private effort by Celera Genomics, culminated in the sequencing of the human genome in 2003.

FEATURES OF INHERITANCE

Discrete inheritance and Mendel's laws

At its most fundamental level, inheritance in organisms occurs by means of discrete traits, called genes. This property was first observed by Gregor Mendel, who studied the segregation of heritable traits in pea plants. In his experiments studying the trait for flower color, Mendel observed that the flowers of each pea plant were either purple or white—and never an intermediate between the two colors. These different, discrete versions of the same gene are called alleles.

In the case of pea plants, each organism has two alleles of each gene, and the plants inherit one allele from each parent. Many organisms, including humans, have this pattern of inheritance. Organisms with two copies of the same allele are called homozygous, while organisms with two different alleles are heterozygous.

The set of alleles for a given organism is called its genotype, while the observable trait the organism has is called its phenotype. When organisms are heterozygous, often one allele is called dominant as its qualities dominate the phenotype of the organism, while the other allele is called recessive as its qualities recede and are not observed. Some alleles do not have complete dominance and instead have incomplete dominance by expressing an intermediate phenotype, or codominance by expressing both alleles at once.

When a pair of organisms reproduce sexually, their offspring randomly inherit one of the two alleles from each parent. These observations of discrete inheritance and the segregation of alleles are collectively known as Mendel's first law or the Law of Segregation.

Notation and diagrams

Geneticists use diagrams and symbols to describe inheritance. A gene is represented by a letter (or letters)—the capitalized letter represents the dominant allele and the recessive is represented by lowercase. Often a "+" symbol is used to mark the usual, non-mutant allele for a gene.

In fertilization and breeding experiments (and especially when discussing Mendel's) the parents are referred to as the "P" generation and the offspring as the "F1" (first filial) generation. When the F1 offspring mate with each other, the offspring are called the "F2" (second filial) generation. One of the common diagrams used to predict the result of cross-breeding is the Punnett square.

When studying human genetic diseases, geneticists often use pedigree charts to represent the inheritance of traits. These charts map the inheritance of a trait in a family tree.

Interactions of multiple genes

Human height is a complex genetic trait. Francis Galton's data from 1889 shows the relationship between offspring height as a function of mean parent height. While correlated, remaining variation in offspring heights indicates environment is also an important factor in this trait.

Organisms have thousands of genes, and in sexually reproducing organisms assortment of these genes are generally independent of each other. This means that the inheritance of an allele for yellow or green pea color is unrelated to the inheritance of alleles for white or purple flowers. This phenomenon, known as "Mendel's second law" or the "Law of independent assortment", means that the alleles of different genes get shuffled between parents to form offspring with many different combinations. (Some genes do not assort independently, demonstrating genetic linkage, a topic discussed later in this article.)

Often different genes can interact in a way that influences the same trait. In the Blue-eyed Mary (*Omphalodes verna*), for example, there exists a gene with alleles that determine the color of flowers: blue or magenta. Another gene, however, controls whether the flowers have color at all: color or white. When a plant has two copies of this white allele, its flowers are white—regardless of whether the first gene has blue or magenta alleles. This interaction between genes is called epistasis, with the second gene epistatic to the first.

Many traits are not discrete features (eg. purple or white flowers) but are instead continuous features (eg. human height and skin color). These complex traits are the product of many genes. The influence of these genes is mediated, to varying degrees, by the environment an organism has experienced. The degree to which an organism's genes contribute to a complex trait is called heritability. Measurement of the heritability of a trait is relative—in a more variable environment, the environment has a bigger influence on the total variation of the trait. For example, human height is a complex trait with a heritability of 89% in the United States. In Nigeria, however, where people experience a more variable access to good nutrition and health care, height has a heritability of only 62%.

MOLECULAR BASIS FOR INHERITANCE

DNA and chromosomes

The molecular basis for genes is deoxyribonucleic acid (DNA). DNA is composed of a chain of nucleotides, of which there are four types: adenine (A), cytosine (C), guanine (G), and thymine (T). Genetic information exists in the sequence of these nucleotides, and genes exist as stretches of sequence along the DNA chain. Viruses are the only exception to this rule—sometimes viruses use the very similar molecule RNA instead of DNA as their genetic material.

DNA normally exists as a double-stranded molecule, coiled into the shape of a double-helix. Each nucleotide in DNA preferentially pairs with its partner nucleotide on the opposite strand: A pairs with T, and C pairs with G. Thus, in its two-stranded form, each strand effectively contains all necessary information, redundant with its partner strand. This structure of DNA is the physical basis for inheritance: DNA replication duplicates the genetic information by splitting the strands and using each strand as a template for synthesis of a new partner strand.

Genes are arranged linearly along long chains of DNA sequence, called chromosomes. In bacteria, each cell has a single circular chromosome, while eukaryotic organisms (which includes plants and animals) have their DNA arranged in multiple linear chromosomes. These DNA strands are often extremely long; the largest human chromosome, for example, is about 247 million base pairs in length. The DNA of a chromosome is associated with structural proteins that organize, compact, and control access to the DNA, forming a material called chromatin; in eukaryotes chromatin is usually composed of nucleosomes, repeating units of DNA wound around a core of histone proteins. The full set of hereditary material in an organism (usually the combined DNA sequences of all chromosomes) is called the genome.

While haploid organisms have only one copy of each chromosome, most animals and many plants are diploid, containing two of each chromosome and thus two copies of every gene. The two alleles for a gene are located on identical loci of sister chromatids, each allele inherited from a different parent.

An exception exists in the sex chromosomes, specialized chromosomes many animals have evolved that play a role in determining the sex of an organism. In humans and other mammals the Y chromosome has very few genes and triggers the development of male sexual characteristics, while the X chromosome is similar to the other chromosomes and contains many genes unrelated to sex determination. Females have two copies of the X chromosome, but males have one Y and only one X chromosome—this difference in X chromosome copy numbers leads to the unusual inheritance patterns of sex linked disorders.

Reproduction

When cells divide, their full genome is copied and each daughter cell inherits one copy. This process, called mitosis, is the simplest form of reproduction and is the basis for asexual reproduction. Asexual reproduction can also occur in multicellular organisms, producing offspring that inherit their genome from a single parent. Offspring that are genetically identical to their parents are called clones.

Eukaryotic organisms often use sexual reproduction to generate offspring that contain a mixture of genetic material inherited from two different parents. The process of sexual reproduction alternates between forms that contain single copies of the genome (haploid) and double copies (diploid). Haploid cells fuse and combine genetic material to create a diploid cell with paired chromosomes. Diploid organisms form haploids by dividing, without replicating their DNA, to create daughter cells that randomly inherit one of each pair of chromosomes. Most animals and many plants are diploid for most of their lifespan, with the haploid form reduced to single cell gametes.

Although they do not use the haploid/diploid method of sexual reproduction, bacteria have many methods of acquiring new genetic information. Some bacteria can undergo conjugation, transferring a small circular piece of DNA to another bacterium. Bacteria can also take up raw DNA fragments found in the environment and integrate them into their genome, a phenomenon known as transformation. This processes result in horizontal gene transfer, transmitting fragments of genetic information between organisms that would otherwise be unrelated.

Recombination and linkage

The diploid nature of chromosomes allows for genes on different chromosomes to assort independently during sexual reproduction, recombining to form new combinations of genes. Genes on the same chromosome would theoretically never recombine, however, were it not for the process of chromosomal crossover. During crossover, chromosomes exchange stretches of DNA, effectively shuffling the gene alleles between the chromosomes. This process of chromosomal crossover generally occurs during meiosis, a series of cell divisions that creates haploid germ cells that later combine with other germ cells to form child organisms.

The probability of chromosomal crossover occurring between two given points on the chromosome is related to the distance between them. For an arbitrarily long distance, the probability of crossover is high enough that the inheritance of the genes is effectively uncorrelated. For genes that are closer together, however, the lower probability of crossover means that the genes demonstrate genetic linkage—alleles for the two genes tend to be inherited together. The amounts of linkage between a series of genes can be combined to form a linear linkage map that roughly describes the arrangement of the genes along the chromosome.

GENE EXPRESSION

Genetic code

Genes generally express their functional effect through the production of proteins, which are complex molecules responsible for most functions in the cell. Proteins are chains of amino acids, and the DNA sequence of a gene (through an RNA intermediate) is used to produce a specific protein sequence. This process begins with the production of an RNA molecule with a sequence matching the gene's DNA sequence, a process called transcription.

This messenger RNA molecule is then used to produce a corresponding amino acid sequence through a process called translation. Each group of three nucleotides in the sequence, called a codon, corresponds to one of the twenty possible amino acids in protein—this correspondence is called the genetic code. The flow of information is unidirectional: information is transferred from nucleotide sequences into the amino acid sequence of proteins, but never from protein back into the sequence of DNA—a phenomenon Francis Crick called the central dogma of molecular biology.

The specific sequence of amino acids results in a unique three-dimensional structure for that protein, and the three-dimensional structures of protein are related to their function. Some are simple structural molecules, like the fibers formed by the protein collagen. Proteins can bind to other proteins and simple molecules, sometimes acting as enzymes by facilitating chemical reactions within the bound molecules (without changing the structure of the protein itself). Protein structure is dynamic; the protein hemoglobin bends into slightly different forms as it facilitates the capture, transport, and release of oxygen molecules within mammalian blood.

A single nucleotide difference within DNA can cause a single change in the amino acid sequence of a protein. Because protein structures are the result of their amino acid sequences, some changes can dramatically change the properties of a protein by destabilizing the structure or changing the surface of the protein in a way that changes its interaction with other proteins and molecules. For example, sickle-cell anemia is a human genetic disease that results from a single base difference within the coding region for the β-globin section of hemoglobin, causing a single amino acid change that changes hemoglobin's physical properties. Sickle-cell versions of hemoglobin stick to themselves, stacking to form fibers that distort the shape of red blood cells carrying the protein. These sickle-shaped cells no longer flow smoothly through blood vessels, having a tendency to clog or degrade, causing the medical problems associated with the disease.

Some genes are transcribed into RNA but are not translated into protein products—these are called non-coding RNA molecules. In some cases these products fold into structures which are involved in critical cell functions (eg. ribosomal RNA and transfer RNA). RNA can also have regulatory effect through hybridization interactions with other RNA molecules (eg. microRNA).

Nature vs. nurture

Although genes contain all the information an organism uses to function, the environment plays an important role in determining the ultimate phenotype—a dichotomy often referred to as "nature vs. nurture." The phenotype of an organism depends on the interaction of genetics with the environment. One example of this is the case of temperature-sensitive mutations. Often, a single amino acid change within the sequence of a protein does not change its behavior and interactions with other molecules, but it does destabilize the structure. In a high temperature environment, where molecules are moving more quickly and hitting each other, this results in the protein losing its structure and failing to function. In a low temperature environment, however, the protein's structure is stable and functions normally. This type of mutation is visible in the coat coloration of Siamese cats, where a mutation in an enzyme responsible for pigment production causes it to destabilize and lose function at high temperatures. The protein remains functional in areas of skin that are colder—legs, ears, tail, and face—and so the cat has dark fur at its extremities.

Environment also plays a dramatic role in effects of the human genetic disease phenylketonuria. The mutation that causes phenylketonuria disrupts the ability of the body to break down the amino acid phenylalanine, causing a toxic build-up of an intermediate molecule that, in turn, causes severe symptoms of progressive mental retardation and seizures. If someone with the phenylketonuria mutation is kept on a strict diet that avoids this amino acid, however, they remain normal and healthy.

Gene regulation

The genome of a given organism contains thousands of genes, but not all these genes need to be active at any given moment. A gene is expressed when it is being transcribed into mRNA (and translated into protein), and there exist many cellular methods of controlling the expression of genes such that proteins are produced only when needed by the cell. Transcription factors are regulatory proteins that bind to the start of genes, either promoting or inhibiting the transcription of the gene. Within the genome of *Escherichia coli* bacteria, for example, there exists a series of genes necessary for the synthesis of the amino acid tryptophan. However, when tryptophan is already available to the cell, these genes for tryptophan synthesis are no longer needed. The presence of tryptophan directly affects the activity of the genes—tryptophan molecules bind to the tryptophan repressor (a transcription factor), changing the repressor's structure such that the repressor binds to the genes. The tryptophan repressor blocks the transcription and expression of the genes, thereby creating negative feedback regulation of the tryptophan synthesis process.

Differences in gene expression are especially clear within multicellular organisms, where cells all contain the same genome but have very different structures and behaviors due to the expression of different sets of genes. All the cells in a multicellular organism derive from a single cell, differentiating into different cell types in response to external and intercellular signals and gradually establishing different patterns of gene expression to create different behaviors. No single gene is responsible for the development of structures within multicellular organisms, these patterns arise from the complex interactions between many cells.

Within eukaryotes there exist structural features of chromatin that influence the transcription of genes, often in the form of modifications to DNA and chromatin that are stably inherited by daughter cells. These features are called "epigenetic" because they exist "on top" of the DNA sequence and retain inheritance from one cell generation to the next. Because of epigenetic features, different cell types grown within the same medium can retain very different properties. Although epigenetic features are generally dynamic over the course of development, some, like the phenomenon of paramutation, have multigenerational inheritance and exist as rare exceptions to the general rule of DNA as the basis for inheritance.

GENETIC CHANGE

Mutations

During the process of DNA replication, errors occasionally occur in the polymerization of the second strand. These errors, called mutations, can have an impact on the phenotype of an organism, especially if they occur within the protein coding sequence of a gene. Error rates are usually very low—1 error in every 10–100 million bases—due to the "proofreading" ability of DNA polymerases. (Without proofreading error rates are a thousand-fold higher; because many viruses rely on DNA and RNA polymerases that lack proofreading ability they experience higher mutation rates.) Processes that increase the rate of changes in DNA are called mutagenic: mutagenic chemicals promote errors in DNA replication, often by interfering with the structure of base-pairing, while UV radiation induces mutations by causing damage to the DNA structure. Chemical damage to DNA occurs naturally as well, and cells use DNA repair mechanisms to repair mismatches and breaks in DNA—nevertheless, the repair sometimes fails to return the DNA to its original sequence.

In organisms that use chromosomal crossover to exchange DNA and recombine genes, errors in alignment during meiosis can also cause mutations. Errors in crossover are especially likely when similar sequences cause partner chromosomes to adopt a mistaken alignment; this makes some regions in genomes more prone to mutating in this way. These errors create large structural changes in DNA sequence—duplications, inversions or deletions of entire regions, or the accidental exchanging of whole parts between different chromosomes (called translocation).

Natural selection and evolution

Mutations produce organisms with different genotypes, and those differences can result in different phenotypes. Many genetic mutations have a negligible effect on an organism's phenotype, health, and reproductive fitness. Mutations that do have an effect are often deleterious, but occasionally mutations arise that are beneficial in the current environmental context of the organism. Population genetics research studies the distributions of these genetic differences within populations and how the distributions change over time. Changes in the frequency of an allele in a population can be influenced by natural selection, where a given allele's higher rate of survival and reproduction causes it to become more frequent in the population over time. Genetic drift can also occur, where chance events lead to random changes in allele frequency.

Over many generations, the genomes of organisms can change, resulting in the phenomenon of evolution. Mutations and the selection for beneficial mutations can cause a species to evolve into forms that better survive their environment, a process called adaptation. New species are formed through the process of speciation, a process often caused by geographical separations that allow different populations to genetically diverge.

As sequences diverge and change during the process of evolution, these differences between sequences can be used as a molecular clock to calculate the evolutionary distance between them. Genetic comparisons are generally considered the most accurate method of characterizing the relatedness between species, an improvement over the sometimes deceptive comparison of phenotypic characteristics. The evolutionary distances between species can be combined to form evolutionary trees—these trees represent the common descent and divergence of species over time, although they cannot represent the transfer of genetic material between unrelated species (known as horizontal gene transfer and most common in bacteria).

RESEARCH AND TECHNOLOGY

Model organisms and genetics

Although geneticists originally studied inheritance in a wide range of organisms, researchers began to specialize in studying the genetics of a particular subset of organisms. The fact that significant research already existed for a given organism would encourage new researchers to choose it for further study, and so eventually a few model organisms became the basis for most genetics research. Common research topics in model organism genetics include the study of gene regulation and the involvement of genes in development and cancer.

Organisms were chosen, in part, for convenience—short generation times and facile genetic manipulation made some organisms popular genetics research tools. Widely used model organisms include the gut bacterium *Escherichia coli*, the plant *Arabidopsis thaliana*, baker's yeast (*Saccharomyces cerevisiae*), the nematode *Caenorhabditis elegans*, the common fruit fly (*Drosophila melanogaster*), and the common house mouse (*Mus musculus*).

Medical genetics research

Medical genetics seeks to understand how genetic variation relates to human health and disease. When searching for an unknown gene that may be involved in a disease, researchers commonly use genetic linkage and genetic pedigree charts to find the location on the genome associated with the disease. At the population level, researchers take advantage of Mendelian randomization to look for locations in the genome that are associated with diseases, a technique especially useful for multigenic traits not clearly defined by a single gene. Once a candidate gene is found, further research is often done on the same gene (called an orthologous gene) in model organisms. In addition to studying genetic diseases, the increased availability of genotyping techniques has led to the field of pharmacogenetics—studying how genotype can affect drug responses.

Although it is not an inherited disease, cancer is also considered a genetic disease. The process of cancer development in the body is a combination of events. Mutations occasionally occur within cells in the body as they divide—while these mutations will not be inherited by any offspring, they can affect the behavior of cells, sometimes causing them to grow and divide more frequently. There are biological mechanisms that attempt to stop this process—signals are given to inappropriately dividing cells that should trigger cell death, but sometimes additional mutations occur that cause cells to ignore these messages. An internal process of natural selection occurs within the body and eventually mutations accumulate within cells to promote their own growth, creating a cancerous tumor that grows and invades various tissues of the body.

Research techniques

DNA can be manipulated in the laboratory. Restriction enzymes are a commonly used enzyme that cuts DNA at specific sequences, producing predictable fragments of DNA. The use of ligation enzymes allows these fragments to be reconnected, and by ligating fragments of DNA together from different sources, researchers can create recombinant DNA. Often associated with genetically modified organisms, recombinant DNA is commonly used in the context of plasmids—short circular DNA fragments with a few genes on them. By inserting plasmids into bacteria and growing those bacteria on plates of agar (to isolate clones of bacteria cells), researchers can clonally amplify the inserted fragment of DNA (a process known as molecular cloning). (Cloning can also refer to the creation of clonal organisms, through various techniques.)

DNA can also be amplified using a procedure called the polymerase chain reaction (PCR). By using specific short sequences of DNA, PCR can isolate and exponentially amplify a targeted region of DNA. Because it can amplify from extremely small amounts of DNA, PCR is also often used to detect the presence of specific DNA sequences.

DNA sequencing and genomics

One of the most fundamental technologies developed to study genetics, DNA sequencing allows researchers to determine the sequence of nucleotides in DNA fragments. Developed in 1977 by Frederick Sanger and coworkers, chain-termination sequencing is now routinely used to sequence DNA fragments. With this technology, researchers have been able to study the molecular sequences associated with many human diseases. As sequencing has become less expensive and with the aid of computational tools, researchers have sequenced the genomes of many organisms by stitching together the sequences of many different fragments (a process called genome assembly). These technologies were used to sequence the human genome, leading to the completion of the Human Genome Project in 2003. The large amount of sequences available has created the field of genomics, research that uses computational tools to search for and analyze patterns in the full genomes of organisms. Genomics can also be considered a subfield of bioinformatics, which uses computational approaches to analyze large sets of biological data.

DNA STRUCTURE AND GENE EXPRESSION

Building blocks and chemical bonds in DNA, RNA and polypeptides

Molecular genetics is primarily concerned with the interrelationship between the information macromolecules DNA (deoxyribonucleic acid) and RNA (ribonucleic acid) and how these molecules are used to synthesize polypeptides, the basic component of all proteins. In some viruses RNA is the hereditary material, but in all cells genetic information is stored in DNA molecules. Selected regions of the cellular DNA molecules serve as templates for synthesizing RNA molecules. The great majority of the RNA molecules are in turn used to specify the synthesis of polypeptides, either directly or by assisting at different stages in gene expression. As the vast majority of gene expression is dedicated to polypeptide synthesis, proteins are the major functional end-points of the DNA template and account for the majority of the dry weight of a cell. The term protein was derived from the Greek *proteios*, meaning 'of the first rank' and reflects the important roles of proteins in diverse cellular functions, as enzymes, receptors, storage proteins, transport proteins, transcription factors, signaling molecules, hormones, etc.

DNA, RNA and polypeptides are large polymers defined by a linear sequence of simple repeating units

In eukaryotes individual DNA molecules are found in the chromosomes of the nucleus and in mitochondria, and also in the chloroplasts of plant cells. They are large polymers, with a linear backbone of alternating sugar and phosphate residues. The sugar in DNA molecules is deoxyribose, a 5 carbon sugar, and successive sugar residues are linked by covalent phosphodiester bonds. Covalently attached to carbon atom number 12 (one *prime*) of each sugar residue is a nitrogenous base. Four types of base are found: adenine (A), cytosine (C), guanine (G) and thymine (T) and they consist of heterocyclic rings of carbon and nitrogen atoms. They can be divided into two classes: purines (A and G) have two joined heterocyclic rings; pyrimidines (C and T) have a single such ring. A sugar with an attached base is called a nucleoside. A nucleoside with a phosphate group attached at carbon atom 5' or 3' constitutes a nucleotide, the basic repeat unit of a DNA strand. The composition of RNA molecules is similar to that of DNA molecules, but differs in that they contain ribose sugar residues in place of deoxyribose and uracil (U) instead of thymine.

BASE	NUCLEOSIDE (= base + sugar)	NUCLEOTIDE (= nucleoside + phosphate)		
	RIBOSE / DEOXYRIBOSE	MONO-PHOSPHATE	DI-PHOSPHATE	TRI-PHOSPHATE
PURINES				
Adenine	Adenosine	AMP (Adenylate)	ADP	ATP
	Deoxyadenosine	dAMP	dADP	dATP
Guanine	Guanosine	GMP (Guanylate)	GDP	GTP
	Deoxyguanosine	dGMP	dGDP	dGTP
PYRIMIDINES				
Cytosine	Cytidine	CMP (Cytidylate)	CDP	CTP
	Deoxycytidine	dCMP	dCDP	dCTP
Thymine	Thymidine	[TMP] (Thymidylate)	[TDP]	[TTP]
	Deoxythymidine	dTMP	dTDP	dTTP
Uracil	Uridine	UMP (Uridylate)	UDP	UTP
	Deoxyuridine	dUMP	dUDP	dUTP

Figure. Common bases found in nucleic acids with corresponding nucleosides and nucleotides. Ribonucleosides and ribonucleotides are shown in blue. Brackets denote nucleotides that are not normally found. Note that nucleotides with a single monophosphate are often described with names where the suffix -ine of the base is replaced by the suffix -ylate, as in the examples given for ribonucleotides.

Proteins are composed of one or more polypeptide molecules which may be modified by the addition of various carbohydrate side chains or other chemical groups. Like DNA and RNA, polypeptide molecules are polymers consisting of a linear sequence of repeating units, in this case amino acids. The latter consist of a positively charged amino group and a negatively charged carboxylic acid (carboxyl) group connected by a central carbon atom to which is attached an identifying side chain. The 20 different amino acids can be grouped into different classes depending on the nature of their side chains. Classification is based as follows:

- Basic amino acids carry a side chain with a net positive charge; an amino (NH_2) group or histidine ring in the side chain acquires a H^+ ion at physiological pH.
- Acidic amino acids carry a side chain with a net negative charge; a carboxyl group (COOH) in the side chain loses a H^+ ion at physiological pH to form COO^-.
- Uncharged polar amino acids are electrically neutral but carry side chains with polar chemical groups, which are distinguished by having fractional electronic charges [e.g. hydroxyl ($O^{\delta-}$-$H^{\delta+}$) groups, sulfhydryl ($S^{\delta-}$-$H^{\delta+}$) groups, etc.]. Like the basic and acidic amino acids, polar amino acids may react with other groups bearing electric charges.
- Nonpolar neutral amino acids are hydrophobic (water-repelling). They often interact with one another and with other hydrophobic groups.

Polypeptides are formed by a condensation reaction between the amino group of one amino acid and the carboxyl group of the next to form a repeating backbone (-NH-CHR-CO-), where the R side chains differ from one amino acid to another.

Covalent bonds confer stability on DNA, RNA and polypeptide molecules, but weaker noncovalent bonds are critically important in permitting intermolecular associations and in stabilizing structure

The stability of the nucleic acid and protein polymers is primarily dependent on the strong covalent bonds that connect the constituent atoms of their linear backbones. In addition to covalent bonds, a number of weak noncovalent bonds are important in interactions between these molecules and between groups within a single nucleic acid or protein molecule. Typically, such noncovalent bonds are weaker than covalent bonds by a factor of more than 10. Unlike covalent bonds whose strength is determined only by the particular atoms involved, however, the strength of noncovalent bonds is also crucially dependent on their aqueous environment. The structure of water is particularly complex, with a rapidly changing network of noncovalent bonding occurring between the individual H_2O molecules. The predominant force in this structure is the hydrogen bond, a weak electrostatic bond formed between a partially positive hydrogen atom and a partially negative atom which, in the case of water molecules, is an oxygen atom.

Charged molecules are highly soluble in water. Because of the phosphate charges present in their component nucleotides, both DNA and RNA are negatively charged (*polyanions*). Depending on their amino acid composition, proteins may carry a net positive charge (basic proteins) or a net negative charge (acidic proteins). The hydrogen bonding potential of water molecules means that molecules with polar groups (including DNA, RNA and proteins) can form multiple interactions with the water molecules, leading to their solubilization. Thus, even electrically neutral proteins are often readily soluble if they contain an appreciable number of charged or neutral polar amino acids. In contrast, membrane-bound proteins are often characterized by a high content of hydrophobic amino acids which are thermodynamically more stable in the hydrophobic environment of a lipid membrane.

Unlike covalent bonds which require considerable energy input to break them, noncovalent bonds are constantly being made and broken at physiological temperatures. As a result, they readily permit reversible and so *transient* molecular interactions, which are essential for biological function. In the case of nucleic acids and proteins, they play a whole variety of key roles, ensuring faithful replication of DNA, transcription of RNA , codon-anticodon recognition, etc. Although individually weak, the combined action of numerous noncovalent bonds can make large contributions to the stability of the structure (conformation) of these molecules and so can be crucially important for specifying the shape of a macromolecule.

DNA structure and replication

The structure of DNA is an antiparallel double helix

The linear backbone of a DNA molecule and of an RNA molecule consists of alternating sugar residues and phosphate groups. In each case, the bond linking an individual sugar residue to the neighbouring sugar residues is a 3' , 5' phosphodiester bond. This means that a phosphate group links carbon atom 3' of a sugar to carbon atom 5' of the neighbouring sugar.

Whereas the RNA molecules within a cell normally exist as single molecules, the structure of DNA is a double helix in which two DNA molecules (DNA strands) are held together by weak hydrogen bonds to form a DNA duplex. Hydrogen bonding occurs between laterally opposed bases, base pairs, of the two strands of the DNA duplex according to Watson-Crick rules: adenine (A) specifically binds to thymine (T) and cytosine (C) specifically binds to guanine (G). As a result, the base composition of DNA from different cellular sources is not random: the amount of adenine equals that of thymine, and the amount of cytosine equals that of guanine. The base composition of DNA can therefore be specified unambiguously by quoting its %GC (= %G + %C) composition. For example, if a source of DNA is quoted as being 42%GC, the base composition can be inferred to be: G, 21%; C, 21%; A, 29%; T, 29%.

DNA can adopt different types of helical structure. A-DNA and B-DNA are both right-handed helices (ones in which the helix spirals in a clockwise direction as it moves away from the observer). They have respectively 11 and 10 base pairs per turn. Z-DNA is a left-handed helix which has 12 base pairs per turn. Under physiological conditions, most of the DNA in a bacterial or eukaryotic genome is of the B-DNA form in which each helical strand has a *pitch* (the distance occupied by a single turn of the helix) of 3.4 nm. As the phosphodiester bonds link carbon atoms number 3' and number 5' of successive sugar residues, one end of each DNA strand, the so-called 5' end, will have a terminal sugar residue in which carbon atom number 5' is not linked to a neighboring sugar residue. The other end is defined as the 3' end because of a similar absence of phosphodiester bonding at carbon atom number 3' of the terminal sugar residue. The two strands of a DNA duplex are said to be antiparallel because they always associate (anneal) in such a way that the 5' $\rightarrow$ 3' direction of one DNA strand is the opposite to that of its partner.

Genetic information is encoded by the linear sequence of bases in the DNA strands (the primary structure). Consequently, two DNA strands of a DNA duplex are said to have complementary sequences (or to exhibit base complementarity) and the sequence of bases of one DNA strand can readily be inferred if the DNA sequence of its complementary strand is already known. It is usual, therefore, to describe a DNA sequence by writing the sequence of bases of one strand only, and in the 5' $\rightarrow$ 3' direction. This is the direction of synthesis of new DNA molecules during DNA replication, and also of the nascent RNA strand produced during transcription. However, when describing the sequence of a DNA region encompassing two neighboring bases (really a dinucleotide) on one DNA strand, it is usual to insert a 'p' to denote a connecting phosphodiester bond [e.g. CpG means that a cytidine is covalently linked to a neighboring guanosine *on the same DNA strand*, while a CG base pair means a cytosine on one DNA strand is hydrogen-bonded to a guanine *on the complementary strand*.

Intermolecular hydrogen bonding also permits RNA-DNA duplexes and double-stranded RNA formation which are important requirements for gene expression. In addition, hydrogen bonding can occur between bases within a single DNA or RNA molecule. Sequences having closely positioned complementary inverted repeats are prone to forming hairpin structures or loops which are stabilized by hydrogen bonding between bases at the neck of the loop. Such structural constraints, which are additional to those imposed by the primary structure, contribute to the secondary structure of the molecule. Certain RNA molecules, such as transfer RNA (tRNA), show particularly high degrees of secondary structure. Note that in the case of hydrogen bonding in RNA-RNA duplexes and also in intramolecular hydrogen bonding within an RNA molecule, GU base pairs are occasionally found. This form of base-pairing is not particularly stable, but does not significantly disrupt the RNA-RNA helix.

DNA replication is semiconservative and synthesis of DNA strands is semidiscontinuous

During the process of DNA synthesis (DNA replication), the two DNA strands of each chromosome are unwound by a helicase enzyme and each DNA strand directs the synthesis of a complementary DNA strand to generate two daughter DNA duplexes, each of which is identical to the parent molecule. As each daughter DNA duplex contains one strand from the parent molecule and one newly synthesized DNA strand, the replication process is described as semiconservative. The enzyme DNA polymerase catalyzes the synthesis of new DNA strands using the four deoxynucleoside triphosphates (dATP, dCTP, dGTP, dTTP) as nucleotide precursors.

DNA replication is initiated at specific points, which have been termed origins of replication. Starting from such an origin, the initiation of DNA replication results in a replication fork, where the parental DNA duplex bifurcates into two daughter DNA duplexes. As the two strands of the parental DNA duplex are antiparallel, but act individually as templates for the synthesis of a complementary antiparallel daughter strand, it follows that the two daughter strands must run in opposite directions [i.e. the direction of chain growth must be 5' $\rightarrow$ 3' for one daughter strand, the leading strand, but 3' $\rightarrow$ 5' for the other daughter strand, the lagging strand.

The reactions catalyzed by DNA polymerases involve addition, to the free 3' hydroxyl group of the growing DNA chain, of a dNMP moiety provided by a dNTP precursor [the two distal phosphate residues of the dNTP - that is, the β and γ residues - are cleaved and the resulting pyrophosphate group (PP_i) discarded]. This requirement introduces an asymmetry into the DNA replication process: only the leading strand will have a free 3' hydroxyl group at the point of bifurcation. This will permit sequential addition of nucleotides and continuous elongation in the same direction in which the replication fork moves. However, synthesis of the lagging strand has to be accomplished as a progressive series of small (typically 100–1000 nucleotides long) fragments, often referred to as Okazaki fragments. As only the leading strand is synthesized continuously, the synthesis of DNA strands is said to be semidiscontinuous. Each fragment of the lagging strand is synthesized in the 5' → 3' direction, which will be in the opposite direction to that in which the replication fork moves. Successively synthesized fragments are covalently joined at their ends using the enzyme DNA ligase so as to ensure chain growth in the direction of movement of the replication fork.

Five classes of mammalian DNA polymerases are known, including a polymerase that is dedicated to replication of the mitochondrial genome. In individual mammalian chromosomes, DNA replication proceeds bidirectionally, to form replication bubbles from multiple initiation points. The distance between adjacent replication origins is about 50–300 kb, a distance which may be significant in chromosome structure. At different origins, DNA replication is initiated at different times in the S phase of the cell cycle but eventually neighboring replication bubbles will fuse. DNA replication is time-consuming: human cells in culture require about 8 hours to complete the process.

RNA transcription and gene expression

The flow of genetic information is almost exclusively one way: DNA → RNA → protein

The expression of genetic information in all cells is very largely a one-way system: DNA specifies the synthesis of RNA and RNA specifies the synthesis of polypeptides, which subsequently form proteins. Because of its universality, the DNA → RNA → polypeptide (protein) flow of genetic information has been described as the central dogma of molecular biology. The first step, the synthesis of RNA using a DNA-dependent RNA polymerase, is described as transcription and occurs in the nucleus of eukaryotic cells and, to a limited extent, in mitochondria and chloroplasts, the only other organelles which have a genetic capacity in addition to the nucleus. The second step, polypeptide synthesis, is described as translation and occurs in ribosomes, large RNA-protein complexes which are found in the cytoplasm and also in mitochondria and chloroplasts. The RNA molecules which specify polypeptide are known as messenger RNA (mRNA). The expression of genetic information follows a colinearity principle: the linear sequence of nucleotides in DNA is decoded to give a linear sequence of nucleotides in RNA which can be decoded in turn in groups of three nucleotides (*codons*) to give a linear sequence of amino acids in the polypeptide product.

Although DNA is the hereditary material in all present-day cells, it is most likely that early in evolution RNA served that function. RNA molecules can, like DNA, undergo self-replication. However, the 22 hydroxyl group on the ribose residues of RNA makes the sugar-phosphate bonds comparatively unstable chemically. In DNA the deoxyribose residues carry only hydrogen atoms at the 22 position and so DNA is much more suited than RNA to being a stable carrier of genetic information. Many different classes of present-day viruses nevertheless have a genome that consists of RNA, not DNA. Retroviruses such as HIV are a subclass of RNA viruses in which the RNA replicates via a DNA intermediate, using a reverse transcriptase, an RNA-dependent DNA polymerase. Recently it has become clear that eukaryotic cells, including mammalian cells, contain nonviral chromosomal DNA sequences which encode cellular reverse transcriptases. Because some nonviral RNA sequences are known to act as templates for cellular DNA synthesis, the principle of unidirectional flow of genetic information is no longer strictly valid.

Only a small fraction of the DNA in complex organisms is expressed to give a protein or RNA product

Only a small proportion of all the DNA in cells is ever transcribed. According to their needs, different cells transcribe different segments of the DNA (transcription units) which are discrete units, spaced irregularly along the DNA sequence. However, the great majority of the cellular DNA is never transcribed in any cell.

Moreover, only a portion of the RNA made by transcription is translated into polypeptide. This is because:

- *some transcription units are expressed to give an RNA molecule other than mRNA and so do not specify polypeptides directly*. Instead, the RNA products are mature molecules which can serve different functions, as in the case of ribosomal RNAs (rRNAs), tRNAs, and diverse small nuclear (sn) and cytoplasmic RNA molecules;
- *the*primary transcript*(initial transcription product) of those transcription units which do encode polypeptides is subject to RNA processing events*. As a result, much of the initial RNA sequence is discarded to give a much smaller mRNA;
- *only a central part of the mature mRNA is translated*; sections of variable length at each end of the mRNA remain untranslated.

In animal cells, DNA is found in both the nucleus and the mitochondria. The mitochondria have, however, only a small fraction of the total cellular DNA and a very limited number of genes; the vast majority of the DNA of a cell is located in the chromosomes of the nucleus.

The fraction of coding DNA in the genomes of complex eukaryotes is rather small. This is partly a result of the noncoding nature of much of the sequence within genes.

Another reason is that a considerable fraction of the genome of complex eukaryotes contains repeated sequences which are nonfunctional or which are not transcribed into RNA. These include defective copies of functional genes (pseudogenes and gene fragments), and highly repetitive noncoding DNA.

Transcription is the process whereby genetic information in some DNA segments (genes) specifies the synthesis of RNA

RNA synthesis is accomplished using an RNA polymerase enzyme, with DNA as a template and ATP, CTP, GTP and UTP as RNA precursors. The RNA is synthesized as a single strand, with the direction of transcription being 5' → 3'. Chain elongation occurs by adding the appropriate ribonucleoside monophosphate residue (AMP, CMP, GMP or UMP) to the free 3' hydroxyl group at the 3' end of the growing RNA chain. Such nucleotides are derived by splitting a pyrophosphate residue (PP_i) from the appropriate ribonucleoside triphosphate (rNTP) precursors.

This means that the nucleotide at the extreme 5' end *(the initiator nucleotide) will differ from all others in the chain by carrying a 5' triphosphate group.*

Normally, only one of the two DNA strands acts as a template for RNA synthesis. During transcription, double-stranded DNA is unwound and the DNA strand which will act as a template for RNA synthesis forms a transient double-stranded RNA-DNA hybrid with the growing RNA chain. As the RNA transcript is complementary to this template strand, the transcript has the same 5' → 3' direction and base sequence (except that U replaces T) as the opposite, nontemplate strand of the double helix.

For this reason the nontemplate strand is often called the sense strand, and the template strand is often called the antisense strand. In documenting gene sequences it is customary to show only the DNA sequence of the sense strand. Orientation of sequences relative to a gene sequence is commonly dictated by the sense strand and by the direction of transcription (e.g. the 5' end of a gene refers to sequences at the 5' end of the sense strand, and sequences upstream or downstream of a gene refer to sequences which flank the gene at the 5' or 3' ends, respectively, of the sense strand).

In eukaryotic cells, three different RNA polymerase molecules are required to synthesize the different classes of RNA. The vast majority of cellular genes encode polypeptides and are transcribed by RNA polymerase II. Increasing importance, however, is being paid to genes which encode RNA as their mature product: functional RNA molecules are now known to play a wide variety of roles, and catalytic functions have been ascribed to some of them.

Transcription of eukaryotic genes requires interaction between cis-acting transcription elements and trans-acting transcription factors

Eukaryotic RNA polymerases cannot initiate transcription by themselves. Instead, combinations of short sequence elements in the immediate vicinity of a gene act as recognition signals for transcription factors to bind to the DNA in order to guide and activate the polymerase. A major group of such short sequence elements is often clustered upstream of the coding sequence of a gene, where they collectively constitute the promoter. After a number of general transcription factors bind to the promoter region, an RNA polymerase binds to the transcription factor complex and is activated to initiate the synthesis of RNA from a unique location. The transcription factors are said to be *trans*-acting, because they are synthesized by genes which are remotely located, and they migrate to their sites of action. In contrast, the promoter elements are *cis*-acting; their function is limited to the DNA duplex on which they reside.

The promoters of many other genes, including housekeeping genes, lack TATA boxes but often have a GC box, containing variants of the consensus sequence GGGCGG. Other common promoter elements include the CAAT box, often at about -80, which is usually the strongest determinant of promoter efficiency. Note, however, that the GC and CAAT boxes appear to be able to function in either orientation, although their sequences are asymmetrical. In addition to general upstream transcription elements which are recognized by ubiquitous transcription factors, more specific recognition elements are known which are recognized by tissue-restricted transcription factors.

Enhancers comprise groups of *cis*-acting short sequence elements, which can enhance the transcriptional activity of specific eukaryotic genes. However, unlike promoter elements whose positions relative to the transcriptional initiation site are relatively constant, enhancers are located a variable, and often considerable, distance from the transcriptional start site, and their function is independent of their orientation. They appear to bind gene regulatory proteins and, subsequently, the DNA between the promoter and enhancer loops out, allowing the proteins bound to the enhancer to interact with the transcription factors bound to the promoter, or with the RNA polymerase. Silencers are equivalent regulatory elements which can inhibit the transcriptional activity of specific genes.

Tissue-specific gene expression involves selective activation of specific genes, and regions of transcriptionally active chromatin adopt an open conformation

The DNA content of a specific type of eukaryotic cell, a myocyte for example, is virtually identical to that of a lymphocyte, hepatocyte or any other type of nucleated cell from the same organism. What makes the different cell types different is the pattern of genes which are expressed in the cell. Some cells, particularly brain cells, express a large number of different genes. In many other cell types, a large fraction of the genes is transcriptionally inactive. Clearly, the genes that are expressed are the ones which define the functions of the cell. Some of these functions are common ones which are essential for general cell functions and are specified by so-called housekeeping genes. The expression of other genes may be largely restricted to a specific cell type (tissue-specific gene expression). Note, however, that even in the case of genes which show considerable tissue specificity in expression, some gene transcripts occur at very low levels in all cell types (illegitimate or ectopic transcription).

The distinction between transcriptionally active and inactive regions of DNA in a cell is reflected in the structure of the associated chromatin. Transcriptionally inactive chromatin generally adopts a highly condensed conformation and is often associated with regions of the genome which undergo late replication during S phase of the cell cycle. It is associated with tight binding by the histone H1 molecule. By contrast, transcriptionally active DNA adopts a more open conformation and is often replicated early in S phase. It is marked by relatively weak binding by histone H1 molecules and extensive acetylation of the four types of nucleosomal histones, i.e. histones H2A,

H2B, H3 and H4. Additionally, in transcriptionally active chromatin the promoter regions of vertebrate genes are generally characterized by absence of methylated cytosines (see below). Transcription factors can displace nucleosomes, and so the open conformation of transcriptionally active chromatin can be distinguished experimentally because it also affords access to nucleases: at very low concentrations, the enzyme DNase I will digest long regions of nucleosome-free DNA. Although the regulatory regions may contain several sequence-specific binding proteins, the open chromatin structure is marked by the presence of DNase I-hypersensitive sites.

RNA transcription and gene expression

The flow of genetic information is almost exclusively one way: DNA → RNA → protein

The expression of genetic information in all cells is very largely a one-way system: DNA specifies the synthesis of RNA and RNA specifies the synthesis of polypeptides, which subsequently form proteins. Because of its universality, the DNA → RNA → polypeptide (protein) flow of genetic information has been described as the central dogma of molecular biology. The first step, the synthesis of RNA using a DNA-dependent RNA polymerase, is described as transcription and occurs in the nucleus of eukaryotic cells and, to a limited extent, in mitochondria and chloroplasts, the only other organelles which have a genetic capacity in addition to the nucleus. The second step, polypeptide synthesis, is described as translation and occurs in ribosomes, large RNA-protein complexes which are found in the cytoplasm and also in mitochondria and chloroplasts. The RNA molecules which specify polypeptide are known as messenger RNA (mRNA). The expression of genetic information follows a colinearity principle: the linear sequence of nucleotides in DNA is decoded to give a linear sequence of nucleotides in RNA which can be decoded in turn in groups of three nucleotides (*codons*) to give a linear sequence of amino acids in the polypeptide product.

Although DNA is the hereditary material in all present-day cells, it is most likely that early in evolution RNA served that function. RNA molecules can, like DNA, undergo self-replication. However, the 22 hydroxyl group on the ribose residues of RNA makes the sugar-phosphate bonds comparatively unstable chemically. In DNA the deoxyribose residues carry only hydrogen atoms at the 22 position and so DNA is much more suited than RNA to being a stable carrier of genetic information. Many different classes of present-day viruses nevertheless have a genome that consists of RNA, not DNA. Retroviruses such as HIV are a subclass of RNA viruses in which the RNA replicates via a DNA intermediate, using a reverse transcriptase, an RNA-dependent DNA polymerase. Recently it has become clear that eukaryotic cells, including mammalian cells, contain nonviral chromosomal DNA sequences which encode cellular reverse transcriptases. Because some nonviral RNA sequences are known to act as templates for cellular DNA synthesis, the principle of unidirectional flow of genetic information is no longer strictly valid.

Only a small fraction of the DNA in complex organisms is expressed to give a protein or RNA product

Only a small proportion of all the DNA in cells is ever transcribed. According to their needs, different cells transcribe different segments of the DNA (transcription units) which are discrete units, spaced irregularly along the DNA sequence. However, the great majority of the cellular DNA is never transcribed in any cell.

Moreover, only a portion of the RNA made by transcription is translated into polypeptide. This is because:

- *some transcription units are expressed to give an RNA molecule other than mRNA and so do not specify polypeptides directly*. Instead, the RNA products are mature molecules which can serve different functions, as in the case of ribosomal RNAs (rRNAs), tRNAs, and diverse small nuclear (sn) and cytoplasmic RNA molecules;
- *the*primary transcript*(initial transcription product) of those transcription units which do encode polypeptides is subject to RNA processing events*. As a result, much

of the initial RNA sequence is discarded to give a much smaller mRNA;

- *only a central part of the mature mRNA is translated*; sections of variable length at each end of the mRNA remain untranslated.

In animal cells, DNA is found in both the nucleus and the mitochondria. The mitochondria have, however, only a small fraction of the total cellular DNA and a very limited number of genes; the vast majority of the DNA of a cell is located in the chromosomes of the nucleus.

The fraction of coding DNA in the genomes of complex eukaryotes is rather small. This is partly a result of the noncoding nature of much of the sequence within genes. Another reason is that a considerable fraction of the genome of complex eukaryotes contains repeated sequences which are nonfunctional or which are not transcribed into RNA. These include defective copies of functional genes (pseudogenes and gene fragments), and highly repetitive noncoding DNA.

Transcription is the process whereby genetic information in some DNA segments (genes) specifies the synthesis of RNA

RNA synthesis is accomplished using an RNA polymerase enzyme, with DNA as a template and ATP, CTP, GTP and UTP as RNA precursors. The RNA is synthesized as a single strand, with the direction of transcription being 5' → 3'. Chain elongation occurs by adding the appropriate ribonucleoside monophosphate residue (AMP, CMP, GMP or UMP) to the free 3' hydroxyl group at the 3' end of the growing RNA chain. Such nucleotides are derived by splitting a pyrophosphate residue (PP_i) from the appropriate ribonucleoside triphosphate (rNTP) precursors. This means that the nucleotide at the extreme 5' end *(the initiator nucleotide) will differ from all others in the chain by carrying a 5' triphosphate group.*

Normally, only one of the two DNA strands acts as a template for RNA synthesis. During transcription, double-stranded DNA is unwound and the DNA strand which will act as a template for RNA synthesis forms a transient double-stranded RNA-

DNA hybrid with the growing RNA chain. As the RNA transcript is complementary to this template strand, the transcript has the same 5' → 3' direction and base sequence (except that U replaces T) as the opposite, nontemplate strand of the double helix. For this reason the nontemplate strand is often called the sense strand, and the template strand is often called the antisense strand. In documenting gene sequences it is customary to show only the DNA sequence of the sense strand. Orientation of sequences relative to a gene sequence is commonly dictated by the sense strand and by the direction of transcription (e.g. the 5' end of a gene refers to sequences at the 5' end of the sense strand, and sequences upstream or downstream of a gene refer to sequences which flank the gene at the 5' or 3' ends, respectively, of the sense strand).

In eukaryotic cells, three different RNA polymerase molecules are required to synthesize the different classes of RNA. The vast majority of cellular genes encode polypeptides and are transcribed by RNA polymerase II. Increasing importance, however, is being paid to genes which encode RNA as their mature product: functional RNA molecules are now known to play a wide variety of roles, and catalytic functions have been ascribed to some of them.

Transcription of eukaryotic genes requires interaction between cis-acting transcription elements and trans-acting transcription factors

Eukaryotic RNA polymerases cannot initiate transcription by themselves. Instead, combinations of short sequence elements in the immediate vicinity of a gene act as recognition signals for transcription factors to bind to the DNA in order to guide and activate the polymerase. A major group of such short sequence elements is often clustered upstream of the coding sequence of a gene, where they collectively constitute the promoter. After a number of general transcription factors bind to the promoter region, an RNA polymerase binds to the transcription factor complex and is activated to initiate the synthesis of RNA from a unique location. The transcription factors are said to be *trans*-acting, because they are synthesized by genes which are remotely located, and they migrate to their sites of action. In contrast, the promoter elements are *cis*-acting; their function is limited to the DNA duplex on which they reside.

In the case of genes which are actively transcribed by RNA polymerase II, either at a specific stage in the cell cycle (e.g. histones) or in specific cell types (e.g. β-globin), the promoter elements always include a TATA box, often TATAAA or a variant, at a position about 25 bp upstream (-25) from the transcriptional start site. Mutation at the TATA element does not prevent initiation of transcription, but does cause the startpoint of transcription to be displaced from the normal position.

The promoters of many other genes, including housekeeping genes, lack TATA boxes but often have a GC box, containing variants of the consensus sequence GGGCGG. Other common promoter elements include the CAAT box, often at about -80, which is usually the strongest determinant of promoter efficiency. Note, however, that the GC and CAAT boxes appear to be able to function in either orientation, although their sequences are asymmetrical. In addition to general upstream transcription elements which are recognized by ubiquitous transcription factors, more specific recognition elements are known which are recognized by tissue-restricted transcription factors.

Enhancers comprise groups of *cis*-acting short sequence elements, which can enhance the transcriptional activity of specific eukaryotic genes. However, unlike promoter elements whose positions relative to the transcriptional initiation site are relatively constant, enhancers are located a variable, and often considerable, distance from the transcriptional start site, and their function is independent of their orientation.

They appear to bind gene regulatory proteins and, subsequently, the DNA between the promoter and enhancer loops out, allowing the proteins bound to the enhancer to interact with the transcription factors bound to the promoter, or with the RNA polymerase. Silencers are equivalent regulatory elements which can inhibit the transcriptional activity of specific genes.

Tissue-specific gene expression involves selective activation of specific genes, and regions of transcriptionally active chromatin adopt an open conformation

The DNA content of a specific type of eukaryotic cell, a myocyte for example, is virtually identical to that of a lymphocyte, hepatocyte or any other type of nucleated cell from the same organism. What makes the different cell types different is the pattern of genes which are expressed in the cell. Some cells, particularly brain cells, express a large number of different genes. In many other cell types, a large fraction of the genes is transcriptionally inactive. Clearly, the genes that are expressed are the ones which define the functions of the cell. Some of these functions are common ones which are essential for general cell functions and are specified by so-called housekeeping genes. The expression of other genes may be largely restricted to a specific cell type (tissue-specific gene expression). Note, however, that even in the case of genes which show considerable tissue specificity in expression, some gene transcripts occur at very low levels in all cell types (illegitimate or ectopic transcription).

The distinction between transcriptionally active and inactive regions of DNA in a cell is reflected in the structure of the associated chromatin. Transcriptionally inactive chromatin generally adopts a highly condensed conformation and is often associated with regions of the genome which undergo late replication during S phase of the cell cycle. It is associated with tight binding by the histone H1 molecule. By contrast, transcriptionally active DNA adopts a more open conformation and is often replicated early in S phase. It is marked by relatively weak binding by histone H1 molecules and extensive acetylation of the four types of nucleosomal histones, i.e. histones H2A, H2B, H3 and H4. Additionally, in transcriptionally active chromatin the promoter regions of vertebrate genes are generally characterized by absence of methylated cytosines (see below). Transcription factors can displace nucleosomes, and so the open conformation of transcriptionally active chromatin can be distinguished experimentally because it also affords access to nucleases: at very low concentrations, the enzyme DNase I will digest long regions of nucleosome-free DNA. Although the regulatory regions may contain several sequence-specific binding proteins, the open chromatin structure is marked by the presence of DNase I-hypersensitive sites.

RNA processing

The RNA transcript of most eukaryotic genes undergoes a series of processing reactions. Often this involves removal of unwanted internal segments and rejoining of the remaining segments (RNA splicing). Additionally, in the case of RNA polymerase II transcripts, a specialized nucleotide linkage (7-methylguanosine triphosphate) is added to the 5' end of the primary transcript (capping), and adenylate (AMP) residues are sequentially added to the 3' end of mRNA to form a poly(A) tail (polyadenylation).

RNA splicing ensures removal of intronic RNA sequences from the primary transcript and fusion of exonic RNA sequences

The coding sequences of most vertebrate genes, both polypeptide-encoding genes and genes encoding RNA molecules other than mRNA, are split into segments (exons) which are separated by noncoding intervening sequences (introns). Transcription involves the production of RNA sequence complementary to the entire length of the gene, encompassing both exons and introns. Often, however, the RNA transcript undergoes RNA splicing, a series of processing reactions whereby the intronic RNA segments are snipped out and discarded and the exonic RNA segments are joined end-to-end (spliced) to give a shorter RNA product.

The mechanism of RNA splicing is dependent upon the identity of the nucleotide sequences at the exon/intron boundaries (splice junctions). In particular, it is critically dependent on what has been called the GT-AG rule: introns almost always start with GT (or really GU at the RNA level) and end with AG.

Although the conserved GT (GU) and AG dinucleotides are crucially important for splicing, they are not by themselves sufficient to signal the presence of an intron. Comparisons of documented sequences have revealed that sequences adjacent to the GT and AG dinucleotides show a considerable degree of conservation. In addition, a third conserved intronic sequence that is known to be functionally important in splicing is the so-called branch site which is usually located very close to the end of the intron, at most 40 nucleotides before the terminal AG dinucleotide.

The splicing mechanism involves the following sequence:

i. cleavage at the 5' splice junction;

ii. nucleolytic attack by the terminal G nucleotide of the splice donor site at the invariant A of the branch site to form a lariat-shaped structure;

iii. cleavage at the 3' splice junction, leading to release of the intronic RNA as a lariat, and splicing of the exonic RNA segments.

The above reactions are mediated by a large RNA-protein complex, the spliceosome, which consists of five types of snRNA (small nuclear RNA) and more than 50 proteins (Stanley and Guthrie, 1998). Each of the snRNA molecules is attached to specific proteins to form snRNP particles and the specificity of the splicing reaction is established by RNA-RNA base-pairing between the RNA transcript and snRNA molecules. In the case of the vast majority of introns in genes which encode polypeptides, the five snRNA species are: U1, U2, U4, U5 and U6 snRNA. The 5' terminus of the U1 snRNA has a sequence UACUUAC which base-pairs with the splice donor consensus (GUAAGUA). After the U1 snRNP has bound, U2 snRNA recognizes the branch site by a similar base-pairing reaction and, subsequently, interaction between U1 snRNP and U2 snRNP brings the two splice junctions close together. Thereafter a multi-snRNP particle, containing U4, U5 and U6 snRNAs, associates with the U1–U2 snRNP complex.

A second type of spliceosome is also known which processes a much rarer class of intron, the so-called AT-AC intron where the conserved GT and AG dinucleotides are replaced by AT and AC respectively. In this case U11 and U12 snRNA replace the functions of U1 and U2. The spliceosome is envisaged to act in a processive manner: once a 5' splice site is recognized, it scans the RNA sequence until it meets the next 3' splice site (which would be signaled as a target by the branch site consensus sequence located just before it). However, the order in which intronic sequences are removed and their flanking exonic sequences spliced is not governed by their linear order in the RNA transcript; instead, the conformation of the RNA is thought to influence the accessibility of 5' splice sites.

Specialized nucleotides are added to the 5' and 3' ends of most RNA polymerase II transcripts

In addition to RNA splicing, RNA polymerase II transcripts are subject to two additional RNA processing events.

Capping

This occurs shortly after transcription. In the case of primary transcripts which will be processed to give mRNA, a methylated nucleoside, 7-methylguanosine (m^7G) is linked to the first 5' nucleotide of the RNA transcript by a special 5' -5' phosphodiester bond. As this bond effectively bridges the 5' carbon of the m^7G residue to the 5' carbon of the first nucleotide, the 5' end is said to be blocked or capped. Transcripts of the snRNA genes are also capped, but their caps may undergo additional modification. The cap has been envisaged to have several possible functions:

- to protect the transcript from 5' × 3' exonuclease attack (decapped mRNA molecules are rapidly degraded);
- to facilitate transport from the nucleus to the cytoplasm;
- to facilitate RNA splicing;
- to play an important role in the attachment of the 40S subunit of the cytoplasmic ribosomes to the mRNA.

Polyadenylation

Transcription by both RNA polymerase I and III is known to stop after the enzyme recognizes a specific transcription termination site. However, identifying possible termination sites for transcription by RNA polymerase II is difficult because the 3' ends of mRNA molecules are determined by a post-transcriptional cleavage reaction. The sequence AAUAAA is a major element that signals 3' cleavage for the vast majority of polymerase II transcripts (*transcripts from histone genes and snRNA genes are notable exceptions*). Cleavage occurs at a specific site located 15–30 nucleotides downstream of the AAUAAA element.

Following the cleavage point, transcription can continue for hundreds or thousands of nucleotides until termination occurs at one of several later sites. Once cleavage has occurred downstream of the AAUAAA element, about 200 adenylate (i.e. AMP) residues are sequentially added in mammalian cells by the enzyme poly(A) polymerase to form a poly(A) tail. The poly(A) tail has been envisaged to have several possible functions:

- to facilitate transport of the mRNA molecules to the cytoplasm;
- to stabilize at least some of the mRNA molecules in the cytoplasm [shortening of poly(A) tracts is associated with mRNA degradation, but some mRNA species (e.g. actin mRNA) remain stable with little or no poly(A)];
- it may facilitate translation by permitting enhanced recognition of the mRNA by the ribosomal machinery.

In the case of histone genes, which are unique in producing mRNA that does not become polyadenylated, termination of transcription also involves 3' cleavage of the primary transcript. This reaction is dependent on secondary structure in the RNA transcript, including a conserved upstream hairpin sequence and a short downstream sequence which base-pairs with a short sequence at the 5' end of the U7 snRNA.

Translation, post-translational processing and protein structure

Translation is the process whereby mRNA is decoded on ribosomes to specify the synthesis of polypeptides

Following post-transcriptional processing, mRNA transcribed from genes in nuclear DNA migrates to the cytoplasm. Here it engages with ribosomes and other components to direct the synthesis of specific polypeptides. The mitochondria also have ribosomes and a limited capacity for protein synthesis.

Only the central segment of a typical eukaryotic mRNA molecule is translated to specify the synthesis of a polypeptide. The flanking sequences, 5' and 3' untranslated regions (5' UTR; 3' UTR), are originally copied from sequence derived from the 5' and 3' terminal exons and, like the 5' cap and 3' poly(A) tail, assist in binding and stabilizing the mRNA on the ribosomes where translation of the central segment occurs.

Ribosomes are large RNA-protein complexes composed of two subunits. In eukaryotes, cytoplasmic ribosomes have a large 60S subunit and a smaller 40S subunit (*S values* are a measure of how fast large molecular structures will sediment in the ultracentrifuge and are governed by both molecular mass and shape). The 60S subunit contains three types of rRNA molecule: 28S rRNA, 5.8S rRNA and 5S rRNA, and about 50 ribosomal proteins. The 40S subunit contains a single 18S rRNA and over 30 ribosomal proteins. Ribosomes provide a structural framework for polypeptide synthesis *in which the RNA components are predominantly responsible for the catalytic function of the ribosome*; the protein components are thought to enhance the function of the rRNA molecules, and a surprising number of them do not appear to be essential for ribosome function.

The assembly of a new polypeptide from its constituent amino acids is governed by a triplet genetic code. Successive groups of three nucleotides (codons) in the linear mRNA sequence are decoded sequentially in order to specify individual amino acids. The decoding process is mediated by a collection of tRNA molecules, to each of which a specific amino acid has been covalently bound (at the free 3' hydroxyl group of the tRNA by a specific amino acyl tRNA synthetase. Different tRNA molecules bind different amino acids.

Each tRNA has a specific trinucleotide sequence, called the anticodon, at a crucially important site located in the center of one arm of the tRNA. This site provides the necessary specificity to interpret the genetic code: for an amino acid to be inserted in the growing polypeptide chain, the relevant codon of the mRNA molecule must be recognized via base-pairing with a suitably complementary anticodon of the appropriate tRNA molecule.

One model of translation envisages that the 40S ribosomal subunit initially recognizes the 5' cap via the participation of proteins that specifically bind to the cap. It then scans along the mRNA until it encounters the initiation codon, which is almost always AUG, specifying methionine (a few cases are known where ACG, CUG or GUG are used instead). Usually, though not always, the first AUG encountered will be the initiation codon. However, the AUG is recognized efficiently as an initiation codon only when it is embedded in a suitable sequence, the optimal being the sequence: GCCPuCC*AUG*G. The most important determinants in this sequence are the G following the AUG codon, and the purine, preferably A, preceding it by three nucleotides (Kozak, 1996). Subsequently, successive amino acids are incorporated into the growing polypeptide chain by a condensation reaction: the amino group of the incoming amino acid reacts with the carboxyl group of the last amino acid to be incorporated, resulting in a peptide bond between successive residues. This is catalyzed by a peptidyl transferase activity which resides in the RNA component of the large ribosomal subunit.

The genetic code is degenerate

The genetic code is a three-letter code. There are four possible bases to choose from at each of the three base positions in a codon. There are, therefore, 64 ($=4^3$) possible codons, but only 20 different amino acids are specified. As a result, the genetic code is said to be *degenerate*: each amino acid is specified on average by about three different codons. Certain amino acids (such as methionine or tryptophan, in the case of the nuclear genetic code) are specified by only a single codon; others, including leucine and serine, are specified by six codons.

Although there are 64 codons, the corresponding number of tRNA molecules with different anticodons is less: just over 30 types of cytoplasmic tRNA and only 22 types of mitochondrial tRNA. The interpretation of all 64 codons on both cytoplasmic and mitochondrial ribosomes is possible because the normal base-pairing rules are relaxed when it comes to codon-anticodon recognition. The wobble hypothesis states that pairing of codon and anticodon follows the normal A-U and G-C rules for the first two base positions in a codon, but that exceptional 'wobbles' occur at the third position and G-U base pairs are also admitted.

Translation continues until a termination codon is encountered (i.e. UAA, UAG or UGA in the case of nuclear-encoded mRNA; UAA, UAG, AGA or AGG in the case of mitochondrial-encoded mRNA; *Figure 1.22*). The backbone of the primary translation product will therefore have at one end a methionine with a free amino group (the N-terminal end) and at the other end an amino acid with a free carboxyl group (the C-terminal end). Note that although codons are translated in a specific translational reading frame, overlapping genes are occasionally found in eukaryotes, in which different translational reading frames are used.

The predominant step in the control of translation is ribosome binding. In addition to the 5' cap, the 5' UTR (often < 200 bp) and 3' UTR (usually very much longer than the 5' UTR) both play critical roles in mRNA recruitment for translation. Several *cis*-acting elements that are involved in this process have been characterized and, in addition, a few *trans*-acting factors which bind to these elements have been identified. It is possible that the 5' and 3' UTR sequences interact to enhance translation. The 3' UTR has a key role in translational regulation and signals for controlling translation, mRNA stability and localization have all been found in this region. The mitochondrial protein synthesis machinery is similar to that found in cytoplasmic ribosomes. For example, mitochondrial ribosomes contain two subunits, a large subunit with a 23S rRNA and several proteins, and a small subunit with a 16S rRNA and several ribosomal proteins. However, the components are assembled from the products of both mitochondrial and nuclear genes: the RNA components are synthesized in the mitochondria (the two rRNA molecules and 22 types of tRNA and a few, usually about 10, types of mRNA.

Post-translational modifications of proteins are frequent and can involve addition of specific chemical groups to specific amino acids and cleavage of the primary translation product

Primary translation products often undergo a variety of modification reactions, involving the addition of chemical groups which are attached covalently to the polypeptide chain at the translational and post-translational levels. This can involve simple chemical modification (hydroxylation, phosphorylation, etc.) of the side chains of single amino acids or the addition of different types of carbohydrate or lipid groups.

Protein modification by addition of carbohydrate groups

Glycoproteins contain oligosaccharides which are covalently attached to the side chains of certain amino acids. Few proteins in the cytosol are *glycosylated*, that is have attached carbohydrate, and those that are carry a single sugar residue, *N*-acetylglucosamine, covalently linked to a serine or threonine residue. By contrast, those proteins which are secreted from cells or exported to lysosomes, the Golgi apparatus or the plasma membrane are glycosylated. Oligosaccharide components of glycoproteins are largely preformed and added *en bloc* to polypeptides. Two major types of glycosylation are recognized:

i. N-glycosylation involves, in most cases, initial transfer of a common oligosaccharide sequence to the side chain NH_2 group of an Asn residue within the endoplasmic reticulum (ER; see *Table 1.6*). Subsequent trimming of residues and replacement with different monosaccharides occurs in the Golgi apparatus;

ii. O-glycosylation

Proteoglycans are proteins with attached *glycosaminoglycans* which usually contain disaccharide repeating units containing glucosamine or galactosamine. The most well characterized proteoglycans are components of the extracellular matrix.

Protein modification by addition of lipid groups

Some proteins, notably membrane proteins, are modified by the addition of fatty acyl or prenyl groups which typically serve as membrane anchors. Examples of fatty acyl groups include the myristoyl group which is a (C_{14}) lipid that is found attached to a glycine residue at the extreme N terminus, and enables the modified protein to interact with a membrane receptor or the lipid bilayer of the membrane. Another fatty acyl group which serves as a membrane anchor is the (C_{16}) palmitoyl group, which becomes attached to the S atom of cysteine residues.

Prenyl groups typically become attached to cysteine residues close to the C terminus, and include farnesyl (C_{15}) groups and geranylgeranyl (C_{20}) groups. Many proteins that participate in signal transduction and protein targeting contain either a farnesyl or a geranylgeranyl unit at their C terminus.

Anchoring of a protein to the outer layer of the plasma membrane uses a different mechanism: the attachment of a glycosylphosphatidyl inositol (GPI) group. This complex glycolipid group contains a fatty acyl group that serves as the membrane anchor which is linked successively to a glycerophosphate unit, an oligosaccharide unit and finally through a phosphoethanolamine unit to the C terminus of the protein. The entire protein except for the GPI anchor is located in the extracellular space.

Post-translational cleavage

The primary translation product may also undergo internal cleavage to generate a smaller mature product. Occasionally, the initiating methionine is cleaved from the primary translation product, as during the synthesis of β-globin. More substantial polypeptide cleavage is observed in the case of the maturation of many proteins, including plasma proteins, polypeptide hormones, neuropeptides, growth factors, etc.

For example, all secreted polypeptides and also polypeptides which are transported across intracellular membranes (e.g. those synthesized in the cytoplasm and transported into the mitochondria) are synthesized initially as precursors, in which a signal sequence (sometimes called a leader sequence) at the N-terminal end acts as a recognition signal for transport across cellular membranes. Thereafter the signal peptide is cleaved from the main polypeptide and degraded. Additionally, in some cases, a single mRNA molecule may specify more than one functional polypeptide chain as a result of proteolytic cleavage of a large precursor polypeptide.

Secretion of proteins and export to specific intracellular locations requires specific localization signals in the coding sequence or specific types of attached group

Proteins synthesized on mitochondrial ribosomes are required to function within the mitochondria. However, the numerous proteins that are synthesized on cytoplasmic ribosomes have diverse functions which may require them to be secreted from the cell where they were synthesized (as with hormones and other intercellular signaling molecules) or to be exported to specific intracellular locations, such as the nucleus (histones, DNA and RNA polymerases, transcription factors, RNA-processing proteins, etc.), the mitochondrion (mitochondrial ribosomal proteins, many respiratory chain components, etc.), peroxisomes, and so on. To do this, a specific localization signal needs to be embedded in the structure of the polypeptide so that it can be sent to the correct address. Usually, the localization signal takes the form of a short peptide sequence. Often, but not always, this constitutes a so-called signal sequence (or leader sequence) which is removed from the protein by a specialized signal peptidase once the sorting process has been achieved.

Signals for export to the endoplasmic reticulum and extracellular space

In the case of secreted proteins the signal peptide comprises the first 20 or so amino acids at the N-terminal end and always includes a substantial number of hydrophobic amino acids. The signal sequence is guided to the ER by a signal recognition particle (SRP), an RNA-protein complex consisting of a small cytoplasmic RNA species, 7SL RNA, and six specific proteins. The SRP complex binds both the growing polypeptide chain and the ribosome and directs them to an SRP receptor protein on the cytosolic side surface of the rough endoplasmic reticulum (RER) membrane. Thereafter a polypeptide can pass into the lumen of the ER, destined for export from the cell, unless there are additional hydrophobic segments which stop the transfer process, as in the case of transmembrane proteins.

Other signals

Like the ER signal, an N-terminal signal sequence is required to traverse the mitochondrial membranes, and is subsequently cleaved. Typically, a mitochondrial signal peptide has, in addition to many hydrophobic amino acids, several positively charged amino acids, usually spaced at intervals of about four amino acids. This structure is thought to form an *amphipathic α-helix*, a helical structure with charged amino acids on one surface and hydrophobic amino acids on another.

Nuclear localization signals can be located just about anywhere within the polypeptide sequence and typically consist of a stretch of four to eight positively charged amino acids, together with neighboring proline residues. Often, however, the signal is *bipartite*, and the positively charged amino acids are found in two blocks of two to four residues, separated by about 10 amino acids. *Note* that some nuclear proteins lack any nuclear localization sequences themselves but are transported into the nucleus with assistance from other nuclear proteins which have appropriate signals.

Lysosomal proteins are targeted to the lysosome by the addition of a mannose 6-phosphate residue which is added in the *cis*-compartment of the Golgi apparatus and is recognized by a receptor protein in the *trans*-compartment of the Golgi.

Protein structure is varied and complex and is not easily predicted from the amino acid sequence of polypeptides

Proteins are composed of one or more polypeptides, each of which can be subject to post-translational modification. They can interact with specific cofactors (for example, divalent cations such as Ca^{2+}, Fe^{2+}, Cu^{2+}, Zn^{2+} or small molecules which are required for functional enzyme activity, e.g. NAD^{+}) or ligands (any molecule which a protein specifically binds), each of which can be powerful influences on the conformation of the protein. At least four different levels of structural organization have been distinguished for proteins.

Within a single polypeptide chain, there is ample scope for hydrogen bonding between different residues; irrespective of the side chains, the oxygen of a peptide bond's carbonyl (CO) group can hydrogen bond to the hydrogen of the NH group of another peptide bond. Fundamental structural units defined by hydrogen bonding between closely neighboring amino acid residues of a single polypeptide include:

- *The α-helix.* This involves formation of a rigid cylinder. The structure is dominated by hydrogen bonding between the carbonyl oxygen of a peptide bond with the hydrogen atom of the amino nitrogen of a peptide bond located four amino acids away. Note that the DNA-binding domains of transcription factors are usually α-helical. An amphipathic α-helix has charged residues on one surface and hydrophobic ones on another. Identical α-helices with a repeating arrangement of nonpolar side chains can coil round each other to form a particular stable structure called a coiled coil. Long rod-like coiled coils are found in many fibrous proteins, such as the α-keratin fibers of skin, hair and nails or fibrinogen of the blood clot.
- *The β-pleated sheet.* This features hydrogen bond formation between opposed peptide bonds in parallel (often really antiparallel) segments of the same polypeptide chain. β-pleated sheets form the core of most, but not all globular proteins.
- *The β-turn.* Hydrogen bonding between the peptide bond CO group of amino acid residue *n* of a polypeptide with the peptide bond NH group of residue *n*+3 results in a hairpin turn. By permitting the polypeptide to reverse direction abruptly, compact globular shapes can be achieved. β-turns are so named because they also often connect antiparallel strands in β-pleated sheets.

More complex structural motifs consisting of combinations of the above structural modules constitute protein domains (compact regions of a protein formed by folding back of the primary structure so that elements of secondary structure can be stacked next to each other). Such domains often represent functional units involved in binding other molecules. In addition, covalent disulfide bridges are often formed between the sulfhydryl (-SH) groups of pairs of cysteine residues occurring within the same polypeptide chain or on different polypeptide chains.

Clearly, the tertiary or quaternary structure of proteins is determined by the primary amino acid sequence. However although secondary structure motifs such as α-helices, β-pleated sheets and β-turns can be predicted by analyzing the primary sequence, the overall three-dimensional structure cannot, at present, be accurately predicted. In addition to the structural complexity of simple polypeptides, many proteins are organized as complex aggregates of multiple polypeptide subunits.

2

CHROMOSOMES IN CELLS

DNA functions in a context. Human DNA is structured into chromosomes that function within cells. Cells divide, differentiate and drive the development of the whole organism. Cell biology and development are huge subjects that we cannot possibly do justice to here, but we start with very brief overviews of these areas to help orient those students who come to human molecular genetics from a nonbiological direction. The main part of this chapter is an introduction to chromosomes - what they are, what they do, and how they can go wrong.

Organization and diversity of cells

All organisms more complex than viruses consist of cells, aqueous compartments bounded by membranes, which under restricted conditions are capable of existing independently. All cells are derived by cell division from other cells. Ultimately, there must be an unbroken chain of cells leading back to the first successful primordial cell that lived maybe 3.5 billion years ago. How that cell formed is an interesting question.

Prokaryotes and eukaryotes represent a fundamental division of living cells

All cellular organisms can be subdivided into two major classes, prokaryotes and eukaryotes, on the basis of the architecture of their cells.

Prokaryotes lack a defined nucleus and have a relatively simple internal organization. Under the electron microscope they appear relatively featureless. They comprise two kingdoms of life: eubacteria which include most of the bacteria; and the archaea, rather poorly understood organisms that superficially resemble bacteria and often grow in unusual environments, such as in acid hot springs, saturated brines, etc. The genome of a prokaryote typically consists of a single small circular chromosome in which the DNA is not packaged in any obviously organized way. Prokaryotes may be simple, but they are not primitive - they have been through far more generations of evolution than we have.

Eukaryotes have a much more complex intracellular organization with internal membranes, membrane-bound organelles including a nucleus, and a well-organized cytoskeleton. Eukaryotic cells have several linear chromosomes in their cell nuclei, in each of which a single very long DNA molecule is elaborately packaged by histone and other proteins. The number and DNA content of the chromosomes vary greatly between species. In general the genome size tends to parallel the complexity of the organism, but there are many exceptions. Most mammals have much the same size genome. Humans do not have specially large genomes, while the cells of an onion and a lily contain respectively about five and 30 times as much DNA as a typical human cell. Eukaryotes are thought to have first apeared about 1.5 billion years ago.

Cell size and shape can vary enormously, but rates of diffusion fix some upper limits

Bacteria and some other simple organisms, such as the yeast *Saccharomyces cerevisiae*, consist of a single cell. Such cells are necessarily able to carry out all the functions that are required to sustain the organism. Multicellular organisms begin life as a single cell but then undergo repeated cell division, cell differentiation and cell turnover. They may end up containing huge numbers of cells. Cell differentiation ensures that individual multicellular organisms are composed of a variety of cell types that can vary greatly in size and shape.

Cells depend on diffusion to coordinate their metabolic activities. As they grow larger, the surface-to-volume ratio decreases. It is thought that the simple internal structure of prokaryotic cells limits their maximum size - typically bacterial cells are 1 ìm in diameter. The complex internal membranes and compartmentalization of eukaryotic cells may be important in allowing them to grow larger. Nevertheless, metabolically active internal regions are seldom more than 15–25 ìm from the cell surface, so that the limit of cell size is typically 30–50 ìm. The average diameter of cells in a multicellular organism falls within the range of 10–30 ìm. Some individual human nerve cells can be as long as 1 metre, but the long projections are very thin. However, the ostrich egg warns us not to over-generalize.

Genetic content of cells: ploidy, the cell cycle and the life cycle

Most cells of humans are diploid. They contain two copies of the human genome. The DNA content and chromosome number of a genome are designated C and n respectively. For humans $C = 3.5 \times 10^{-12}$ g, approximately, and $n = 23$. The DNA content of diploid cells is $2C$ and they have $2n$ chromosomes. Almost all mammals are diploid, but among other organisms there are many examples of species that are normally haploid (n chromosomes, DNA content C), tetraploid ($4n$) or polyploid. Triploidy ($3n$) is less common because triploids have problems with meiosis.

The diploid cells of our body are derived from the original diploid fertilized egg by repeated rounds of mitotic cell division. Each round can be summarized as one turn of the cell cycle. This comprises a short stage of cell division, the M phase and a long intervening interphase. Interphase can be divided into S phase (DNA synthesis), G_1 phase (gap between M phase and S phase) and G_2 phase (gap between S phase and M phase). From anaphase of mitosis right through until DNA duplication in S phase, a chromosome of a diploid cell contains a single DNA double helix and the total DNA content is $2C$. G_1 is the normal state of a cell, and the long-term end state of nondividing cells. Cells enter S phase only if they are committed to mitosis; nondividing cells remain in a modified G_1 stage, sometimes called G_0. The cell cycle diagram can give the impression that all the interesting action happens in S and M phases - but this is an illusion. A cell spends most of its life in G_0 or G_1 phase, and that is where the genome does most of its work. A subset of the diploid body cells constitute the germ line. These give rise to specialized diploid cells in the ovary and testis that can divide by meiosis to produce haploid gametes (sperm and egg). In humans ($n = 23$) each gamete contains 22 autosomes (nonsex chromosomes) plus one sex chromosome. In eggs the sex chromosome is always an X; in sperm it may be an X or a Y. After fertilization the zygote is diploid ($2n$) with the chromosome constitution 46,XX or 46,XY. The other cells of the body, apart from the germline, are known as somatic cells. Somatic cells have no input to succeeding generations - unless animal cloners intervene.

Although most somatic cells are diploid, there are exceptions. Some terminally differentiated cells, red blood cells for example, have no nuclei and others, such as keratinocytes, are devoid of organelles altogether. Such cells are nulliploid. Other cells are polyploid as a result of DNA replication without cell division (endomitosis). Regenerating cells of the liver and other tissues are naturally tetraploid because of endomitosis, while the giant megakaryocytes of the bone marrow usually contain $8C$, $16C$ or $32C$, and individually give rise to thousands of nulliploid platelet cells. Other naturally occurring cells have multiple diploid nuclei as a result of cell fusion.

Development

The development of any animal from a single fertilized egg cell is vastly complicated, but the early stages are common to all animals, and at the molecular level development is controlled by a limited repertoire of developmental programs. All development depends on the basic processes of cell division, differentiation, morphogenesis and programed cell death (apoptosis). Differentiation is driven by gene switching: the difference between one cell type and another is primarily in the range of genes that are active in each cell. Morphogenesis, too, is ultimately driven by gene switching, as particular cells develop the capacity to respond to signals from neighboring cells by moving, dividing or dying. Apoptosis is an integral part of development: cells do not just happen to die, they have an inbuilt death program that is triggered in response to external or internal signals. All these developmental programs depend on cascades of signals and responses that have been remarkably highly conserved throughout the animal kingdom. Unraveling these programs is a major part of biological research. Probably the best introduction to how this is done, and why the results matter, is the book by Lawrence (Further reading).

Only a small percentage of the cells in the early embryo give rise to the mature organism

The early stages of human development are largely concerned with establishing the future placenta and extraembryonic membranes, and implanting the conceptus in the uterine endometrium. About 2 weeks of development and many rounds of mitosis pass before the group of cells that will give rise to the embryo has any separate identity. Embryonic development proper begins with gastrulation in the third week after fertilization.

Cells differ in their potential to divide and differentiate

Multicellular animals begin life as a single cell following fertilization of an egg cell (*oocyte*) by a sperm cell. The fertilized egg proceeds to undergo a series of cell divisions. At the early stages of development, individual cells in the embryo are totipotent: each cell retains the capacity to differentiate into all the different types of cell in the body. As development proceeds cells become more *restricted* in their capacity to generate different types of descendant cells and are said to be pluripotent. Progenitor cells that can only develop into a single cell type are unipotent cells. The processes of cell differentation lead to individual cells acquiring specialized forms and functions.

Histology textbooks recognize about 210 different types of cell in the human body. Some, such as epithelial cells and fibroblasts, perform essentially the same role in a variety of organ systems. Others, such as hepatocytes, may be restricted to an individual organ. Some mature *terminally differentiated* cells do not undergo cell division. Other cells (often distinguished by the suffix *-blast*, as in osteoblasts, chondroblasts, myoblasts, etc.) divide actively and act as precursors of terminally differentiated cells. In some cases, the precursor cells are also capable of undergoing self-renewal and are known as stem cells. Stem cells are an important target of gene therapy.

X inactivation is a special feature of female development

Chromosomally-based sex-determination systems, like the human XX/XY system, create a problem for development. Having the wrong number of chromosomes almost always makes an organism develop abnormally, and yet the two sexes must develop normally with different chromosome constitutions. In humans and other mammals the solution is different for the X and the Y chromosomes. The Y chromosome contains very few genes, and these are mostly genes governing male sexual function, so that females can get by perfectly well without a Y chromosome. The X chromosome, however, contains many genes that play vital roles in both sexes, and so some method of dosage compensation is required, to ensure that cells function normally with either one or two X chromosomes.

Mammals achieve X-chromosome dosage compensation by the mechanism of X-inactivation, often called lyonization, after Dr Mary Lyon who first suggested this mechanism. Early in embryonic development (at late blastocyst stage in the mouse, and probably also in humans), cells somehow count their X chromosomes and then inactivate all but one of them. 46,XY cells (and also 45,X cells) leave their single X active; 46,XX cells (and also 47,XXY cells) inactivate one X, while a 47,XXX cell would inactivate two Xs. Thus, regardless of the number of X chromosomes in the karyotype, there is only one set of active X-linked genes in a cell. Males are constitutionally hemizygous for X chromosome genes (that is, they have only a single copy of each gene), while females become functionally hemizygous: they have only a single functional copy of each gene.

When chromosomes of a female cell are observed at metaphase of mitosis, the active and inactive Xs look the same - but this is because *all* chromosomes at metaphase of mitosis are condensed and largely inactive. After the end of cell division, the inactive X remains condensed while the other chromosomes decondense and resume transcriptional activity. In some cells the inactive X can be seen as a Barr body or sex chromatin body near the membrane of the interphase nucleus. This allows a simple but not very reliable method of sexing interphase cells.

Which X in a 46,XX cell is inactivated is random (with a few exceptions discussed later), so that in a female embryo, some cells will inactivate the paternal X and some the maternal X. Once the choice is made, it is remembered. When the cell divides, the daughter cells inactivate the same X as the mother cell. An adult female is a mosaic of clones derived from different embryonic cells. Within a clone, all the cells inactivate the same X, but between clones the choice is random. If she happens to be a carrier of an X-linked recessive disease, this can have major implications.

X inactivation is a fascinating and imperfectly understood phenomenon. One would like to know how the cell counts its X chromosomes, how the inactivation works, how it is perpetuated, and how it is reversed during gametogenesis. We know that there is not blanket inactivation of all genes on the inactive X; certain regions, and certain genes in other regions, escape inactivation.

Structure and function of chromosomes

Chromosomes as seen under the microscope and illustrated in textbooks are rather misleading. When we look at chromosomes in a dividing cell we see the genome of the cell largely switched off and packed up into neat bundles ready for cell division. The processes of cell division are fascinating in their own right, and errors in packaging or dividing up the genome have major medical consequences. However, it is important to remember that the switched on, functioning interphase chromosome that controls cellular activities is a much more extended and diffuse structure than the metaphase chromosomes. Importantly, it comprises only a single chromatid and one DNA double helix, not the two-chromatid structure of mitotic chromosomes. As functioning organelles, eukaryotic chromosomes seem to require only three classes of DNA sequence element: centromeres, telomeres and origins of replication. This simple requirement has been verified by the successful construction of artificial chromosomes in yeast: large foreign

DNA fragments behave as autonomous chromosomes when ligated to short sequences that specify a functional centromere, two telomeres and a replication origin. Recently mammalian artificial chromosomes have been constructed on similar principles.

Packaging of DNA into chromosomes requires multiple hierarchies of DNA folding

In the cell the structure of each chromosome is highly ordered (Manuelidis, 1990). Even in the interphase nucleus the 2 nm DNA double helix is subject to at least two levels of coiling.

- The most fundamental unit of packaging is the nucleosome. This consists of a central core of eight histone proteins, small highly conserved basic proteins of 102–135 amino acids. Each core comprises two molecules each of histones H2A, H2B, H3 and H4, around which a stretch of 146 bp of double-stranded DNA is coiled in 1.75 turns. Adjacent nucleosomes are connected by a short length of spacer DNA. Electron micrographs of suitable preparations show a 'string of beads' appearance.
- The string of beads, approximately 10 nm in diameter, is in turn coiled into a chromatin fiber of 30 nm diameter. The interphase chromosome seems to consist of these chromatin fibers, probably organized into long loops as described below.

During cell division, the chromosomes become ever more highly condensed. The DNA in a metaphase chromosome is compacted to about 1/10 000 of its stretched-out length. Loops of the 30 nm chromatin fiber, containing 20–100 kb of DNA per loop, are attached to a central *scaffold*. This consists of nonhistone acidic proteins, notably topoisomerase II, an enzyme which has the interesting ability to pass one DNA double helix through another by cutting a gap and repairing it. Topoisomerase II and some other chromatin proteins are known to bind to AT-rich sequences, and the chromatin loops may be attached by stretches of several hundred base pairs of highly AT-rich (>65%) DNA (scaffold attachment regions). In the chromatids of a metaphase chromosome the loop-scaffold complex is compacted yet further by coiling.

Chromosomes as functioning organelles: the centromere

Normal chromosomes have a single centromere that is seen under the microscope as the primary constriction, the region at which sister chromatids are joined. The centromere is essential for segregation during cell division. Chromosome fragments that lack a centromere (acentric fragments) do not become attached to the spindle, and so fail to be included in the nuclei of either of the daughter cells.

During late prophase of mitosis, a pair of kinetochores forms at each centromere, one attached to each sister chromatid. Multiple microtubules attach to each kinetochore, linking the centromere of a chromosome and the two spindle poles. At anaphase, the kinetochore microtubules pull the two sister chromatids toward opposite poles of the spindle. Kinetochores play a central role in this process, by controlling assembly and disassembly of the attached microtubules and, through the presence of motor molecules, by ultimately driving chromosome movement.

Specific DNA sequences presumably specify the structure and function of centromeres. In simple eukaryotes, the sequences that specify centromere function are very short. For example, in the yeast *Saccharomyces cerevisiae* the centromere element (CEN) is about 110 bp long, comprising two highly conserved flanking elements of 9 bp and 11 bp and a central AT-rich segment of about 80–90 bp. The centromeres of such cells are interchangeable - a CEN fragment derived from one yeast chromosome can replace the centromere of another with no apparent consequence. In mammals, centromeres comprise hundreds of kilobases of repetitive DNA, some nonspecific and some chromosome-specific.

Chromosomes as functioning organelles: origins of replication

The DNA in most diploid cells normally replicates only once per cell cycle. The initiation of replication is controlled by *cis*-acting sequences that lie close to the points at which DNA synthesis is initiated. Probably these are sites at which *trans*-acting proteins bind. Eukaryotic origins of replication have been most comprehensively studied in yeast, where the presence of a putative replication origin can be tested by a genetic assay. To test the ability of a random fragment of yeast DNA to promote autonomous replication, it is incorporated into a bacterial plasmid together with a yeast gene that is essential for growth of yeast cells.

This construct is used to transform a mutant yeast that lacks the essential gene. The transformed cells can only form colonies if the plasmid can replicate in yeast cells. However, the bacterial replication origin in the plasmid does not function in yeast, therefore the few plasmids that transform at high efficiency must possess a sequence within the inserted yeast fragment that confers the ability to replicate extra-chromosomally at high efficiency - that is an autonomously replicating sequence (ARS) element.

ARS elements are thought to derive from authentic origins of replication and, in some cases, this has been confirmed by mapping a specific ARS element to a specific chromosomal location and demonstrating that DNA replication is indeed initiated at this location. ARS elements extend for only about 50 bp and consist of an AT-rich region which contains a conserved core consensus and some imperfect copies of this sequence. In addition, the ARS elements contain a binding site for a transcription factor and a multiprotein complex is known to bind to the origin.

Mammalian replication origins have been much less well defined because of the absence of a genetic assay. Some initiation sites have been studied, but such studies have not been able to identify a unique origin of replication. This has led to speculation that replication can be initiated at multiple sites over regions tens of kilobases long. Mammalian artificial chromosomes seem to work without specific ARS sequences being provided. Computer analysis of regions encompassing several eukaryotic origins of replication, including some human and other mammalian examples, identified a consensus DNA sequence WAWTTDDWWWDHWGWHMAWTT where W = A or T; D = A or G or T; H = A or C or T; and M = A or C (Dobbs *et al.*, 1994).

Chromosomes as functioning organelles: the telomeres

Telomeres are specialized structures, comprising DNA and protein, which cap the ends of eukaryotic chromosomes. They have several likely functions:

- Maintaining the structural integrity of a chromosome. If a telomere is lost, the resulting chromosome end is unstable. It has a tendency either to fuse with the ends of other broken chromosomes, to be involved in recombination events or to be degraded. The loop structure of human telomeres (see below) means that natural chromosomes have no free DNA end.
- Ensuring complete replication of the extreme ends of chromosomes. During DNA replication, synthesis of the lagging strand is discontinuous and requires the presence of some DNA ahead of the sequence which is to be copied to serve as the template for an RNA primer. However, at the extreme end of a linear molecule, there can never be such a template, and a different mechanism is required to solve the problem of replicating the ends of a linear DNA molecule.
- Helping establish the three-dimensional architecture of the nucleus and/or chromosome pairing. Chromosome ends appear to be tethered to the nuclear membrane, suggesting that telomeres help position chromosomes.

Eukaryotic telomeres consist of a long array of tandem repeats. One DNA strand contains TG-rich sequences and terminates in the 3' end; the complementary strand is CA-rich. Unlike centromeres, the sequence of telomeres has been highly conserved in evolution - there is considerable similarity in the simple sequence repeat, for example TTGGGG (*Paramecium*), TAGGG (*Trypanosoma*) TTTAGGG (*Arabidopsis*) and TTAGGG (*Homo sapiens*).

The problem of replicating the ends of a chromosome has been solved by extending the synthesis of the leading strand using a specialized enzyme, telomerase. This RNA-protein complex carries within its RNA component a short sequence which will act as a template to prime extended DNA synthesis of telomeric DNA sequences on the leading strand. Further extension of the leading strand provides the necessary template for DNA polymerase α to complete synthesis of the lagging strand. This mechanism leaves the telomere itself with a protruding 3' end. In mammalian chromosomes, the single-stranded end is believed to loop round and invade the double helix several kilobases proximally, producing a triple-stranded structure resembling the mitochondrial D-loop, which is stabilized by binding telomere-specific proteins (Greider, 1999). However, the actual nature of the telomere sequence may not be important. The telomere length is known to be highly variable and is subject to genetic control.

Just internal to the essential telomeric repeats, eukaryotic chromosomes also have a more complex set of repeats called subtelomeric or telomere-associated repeats. Their sequences are not conserved in eukaryotes and their function is unknown.

Heterochromatin and euchromatin

In the interphase nucleus most of the chromatin (euchromatin) exists in an extended state, dispersed through the nucleus and staining diffusely. However, some chromatin remains highly condensed throughout the cell cycle and forms dark-staining regions (heterochromatin). Genes located in euchromatin may or may not be expressed, depending on the cell type and its metabolic requirements, but genes that are located within heterochromatin, either naturally or as the result of a chromosomal rearrangement, are very unlikely to be expressed. There are two classes of heterochromatin:

- Constitutive heterochromatin is always inactive and condensed. It consists largely of repetitive DNA and is found in and around the centromeres of chromosomes and in certain other regions.
- Facultative heterochromatin can exist in either a genetically active (decondensed) or an inactive and condensed form, as in the case of mammalian X-chromosome inactivation.

In euchromatin, the G bands partake of some of the properties of heterochromatin, but to a lesser degree. G band chromatin in metaphase chromosomes is more condensed than R band chromatin, and data on CpG island distribution show that G bands are relatively poor in genes. The subset of R bands that are revealed by T-banding have a particularly high density of genes.

Mitosis and meiosis are the two types of cell division

Mitosis is the normal form of cell division

As a person develops from an embryo, through fetus and infant to an adult, cell divisions are needed to generate the large numbers of cells required. Additionally, many cells have a limited lifespan, so there is a continuous requirement to generate new cells in the adult. All these cell divisions occur by mitosis. Mitosis is the normal process of cell division, from cleavage of the zygote to death of the person. In the lifetime of a human there may be something like 10^{17} mitotic divisions.

The M phase of the cell cycle consists of the various stages of nuclear division (prophase, prometaphase, metaphase, anaphase and telophase of mitosis), and cell division (cytokinesis), which overlaps the final stages of mitosis. In preparation for cell division, the previously highly extended chromosomes contract and condense so that, by metaphase of mitosis, they are readily visible under the microscope. Even though the DNA was replicated some time previously, it is only at prometaphase that individual chromosomes can be seen to comprise two sister chromatids, attached at the centromere.

The mitotic spindle is formed from tubulin-based microtubules and microtubule-associated proteins. Polar fibers, which extend from the two poles of the spindle towards the equator, develop at prophase while the nuclear membrane is still intact. Kinetochore fibers do not develop until prometaphase. These fibers attach to the kinetochore, a large multiprotein structure attached to the centromere of each chromatid, and extend in the direction of the spindle poles. The interaction between the different spindle fibers pulls the chromosomes towards the center, and by metaphase each chromosome is independently aligned on the equatorial plane (metaphase plate). Paternal and maternal homologs do not associate at all during mitosis. Following centromere division at anaphase, the spindle fibers pull the separated sister chromatids of each chromosome to opposite poles. The DNA of the two sister chromatids is identical, barring any errors in DNA replication. Thus the effect of mitosis is to generate daughter cells that contain precisely the same DNA sequences.

Meiosis is a specialized form of cell division giving rise to sperm and egg cells

Primordial germ cells migrate into the embryonic gonad and engage in repeated rounds of mitosis many more in males than in females, which may be a significant factor in explaining sex differences in mutation rates to form oogonia in females and spermatogonia in males. Further growth and differentiation produces primary oocytes in the ovary and primary spermatocytes in the testis. These specialized diploid cells can undergo meiosis. Meiosis involves two successive cell divisions but only one round of DNA replication, so the products are haploid. In males, the product is four spermatozoa; in females, however, the cytoplasm divides unequally at each stage: the products of meiosis I (the first meiotic division) are a large secondary oocyte and a small cell (polar body). The secondary oocyte then gives rise to the large mature egg cell and a second polar body.

There are two crucial differences between mitosis and meiosis.

- The products of mitosis are diploid; the products of meiosis are haploid.
- The products of mitosis are genetically identical; the products of meiosis are genetically different.

Mitosis involves a single turn of the cell cycle. The DNA is replicated in S phase and the two copies are divided exactly equally between the daughter cells in M phase. Meiosis is also preceded by one round of DNA synthesis, but then there are two cell divisions without intervening DNA synthesis, so that the products end up haploid. The second division of meiosis is identical to mitosis, but the first division has important differences whose purpose is to generate genetic diversity between the daughter cells. This is done by two mechanisms, independent assortment of paternal and maternal homologs, and recombination.

Independent assortment of paternal and maternal homologs

During meiosis I the maternal and paternal homologs of each chromosome pair form a bivalent by pairing together (synapsis). Each chromosome consists of two sister chromatids following DNA replication, so that the bivalent is a four-stranded structure at the metaphase plate. Spindle fibers then pull one complete chromosome (two chromatids) to either pole. However, for each of the 23 homologous pairs, the choice of which homolog enters which daughter cell is independent. This allows 2^{23} or about 8.4×10^6 possible combinations of parental chromosomes to be produced by one person.

Recombination

During prophase of meiosis I the synapsed homologs within each bivalent exchange segments in a random way. At the zygotene stage, each pair of homologs begins to form a synaptonemal complex consisting of the two chromosomes in close apposition, separated by a long linear protein core. Completion of this complex marks the start of the pachytene stage, which is when recombination (or crossover) occurs. Crossing-over involves physical breakage of the double helix in one paternal and one maternal chromatid, and joining of maternal and paternal ends. Overall, the combination of recombination between homologs in prophase I plus independent assortment of homologs at anaphase I ensures that a single individual can produce an almost unlimited number of genetically different gametes.

The mechanism allowing alignment of the homologs is not understood. However, it is thought that such close apposition is required for recombination. Recombination nodules, very large multiprotein assemblies located at intervals on the synaptonemal complex, are thought to mediate the recombination events. The two homologs can be seen to be physically connected at specific points. Each such connection is described as a chiasma (plural chiasmata) and marks a crossover point. There are an average of 55 chiasmata in a male meiotic cell, and maybe 50% more in female meiosis.

In addition to their role in recombination, chiasmata are thought to be essential for correct chromosome segregation at meiosis I. By holding the maternal and paternal homologs of each chromosome pair together on the spindle until anaphase I, they have a role that is analogous to that of the centromeres in mitosis and meiosis II. There is genetic evidence that children with wrong numbers of chromosomes are often the product of gametes where a bivalent lacked crossovers.Meiosis II appears identical to mitosis, except that there are only 23 chromosomes instead of 46. Each chromosome consists of two chromatids, and these are separated in anaphase II. However, there is one difference. The sister chromatids of a mitotic chromosome are identical, being copies of each other. The two chromatids of a chromosome in meiosis II may be genetically different as a result of crossovers in meiosis I.

X-Y pairing and the pseudoautosomal regions

In female meiosis, each chromosome has a fully homologous partner, and the two Xs synapse and cross over just like any other pair of homologs. In male meiosis there is a problem. The human X and Y sex chromosomes are very different from one another. Nevertheless, they do pair in prophase I in males, thus ensuring that at anaphase I each daughter cell receives one sex chromosome, either the X or the Y. X-Y pairing is end-to-end rather than along the whole length, and it is made possible by a 2.6 Mb region of homology between the X and Y chromosomes at the tips of their short arms. Pairing is sustained by an obligatory crossover in this region. Genes in the pairing segment have some interesting properties:

- they are present as homologous copies on the X and Y chromosomes;
- they are not subject to X-inactivation (as expected since each sex has two copies);

- because of the crossing over, alleles at these loci do not show the normal X-linked or Y-linked patterns of inheritance, but segregate like autosomal alleles.

Because of this behavior, this region is known as the major pseudoautosomal region. A second smaller pseudoautosomal region of 320 kb is located at the tips of the long arms of both chromosomes, but pairing and crossing-over in this minor pseudoautosomal region is not an obligatory feature of male meiosis.

Studying human chromosomes

Mitotic chromosomes can be seen in any dividing cell, but most conveniently in lymphocytes; meiotic chromosomes are hard to study in humans

Chromosomes can only be seen in dividing cells, and obtaining dividing cells directly from the human body is difficult. Bone marrow is a possible source, but it is much easier all round to take an accessible source of nondividing cells and culture them in the laboratory. Blood is the material of choice - most people don't mind giving a few millilitres, and the T lymphocytes in blood can be easily induced to divide by treatment with lectins such as phytohemagglutinin. Other common sources include fibroblasts grown from skin biopsies, and (for prenatal diagnosis) chorionic villi or fetal cells shed into the amniotic fluid.

Although chromosomes were described accurately in some organisms as early as the 1880s, for many decades all attempts to prepare spreads of human chromosomes produced a tangle that defied analysis. The key to getting analyzable spreads was a new technique, growing cells in liquid suspension and treating them with hypotonic saline to make them swell. This allowed the first good quality preparations to be made in 1956. White cells from blood are put into a rich culture medium laced with phytohemagglutinin and allowed to grow for 48–72 hours, by which time they should be dividing freely. Nevertheless, because M phase occupies only a small part of the cell cycle, few cells will be actually dividing at any one time. The mitotic index (proportion of cells in mitosis) is increased by treating the culture with a spindle disrupting agent such as colcemid. Cells reach M phase of the cycle, but are unable to leave it, and so cells accumulate in metaphase of mitosis.

Often it is preferable to study prometaphase chromosomes, which are less contracted and so show more detail. Cell cultures can be prevented from cycling by thymidine starvation; when the block is released the cells progress through the cycle synchronously. By trial and error, the time after release can be determined when a good proportion of cells are in the desired prometaphase stage.

Meiosis can only be studied in testicular or ovarian samples. Female meiosis is especially difficult, as it is active only in fetal ovaries, whereas male meiosis can be studied in a testicular biopsy from any post-pubertal male who is willing to give one. The results of meiosis can be studied by analyzing chromosomes from sperm, although the methodology for this is cumbersome. Meiotic analysis is used for some investigations of male infertility.

Chromosomes are identified by their size, centromere position and banding pattern

Until the 1970s chromosomes were identified on the basis of their size and the position of the centromeres. This allowed chromosomes to be classified into groups but not unambiguously identified. The introduction of banding techniques finally allowed each chromosome to be identified, as well as permitting more accurate definition of translocation breakpoints, subchromosomal deletions, etc. Banding resolution can be increased by using more elongated chromosomes, for example chromosomes from prometaphase or earlier, rather than metaphase. Typical high-resolution banding procedures for human chromosomes can resolve a total of 400, 550 or 850 bands.The chromosome constitution is described by a karyotype that states the total number of chromosomes and the sex chromosome constitution. Human females and males are 46,XX and 46,XY respectively. When there is a chromosomal abnormality the karyotype also describes the type of abnormality and the chromosome bands or sub-bands affected.

Chromosome banding picks out structural organization on a 1–10 Mb scale

Various treatments involving denaturation and/or enzymatic digestion, followed by incorporation of a DNA-specific dye, can cause human and other mitotic chromosomes to stain as a series of light and dark bands. Banding patterns are interesting (as well as being useful to cytogeneticists) because they provide evidence of some sort of structure over 1–10 Mb regions. The banding patterns correlate with other properties. Regions that stain as dark G bands replicate late in S phase of the cell cycle and contain more condensed chromatin, while R bands (light G bands) generally replicate early in S phase, and have less condensed chromatin. Genes are mostly concentrated in the R bands, while the later replicating, more condensed G-band DNA is less active transcriptionally. There are also differences in the types of dispersed repeat elements found in G and R bands.

Bands similar to G bands can be produced by staining with quinacrine, which preferentially binds to AT-rich DNA, while the R-banding pattern can be elicited using chromomycin, which preferentially binds GC-rich DNA. However, the AT content of human G band DNA is only a few per cent higher than R band DNA. Saitoh and Laemmli (1994) suggested the difference depends on differences in the scaffold-loop structure. Chromatin loops are thought to attach to the chromosome scaffold at special scaffold attachment regions (SARs). According to the Saitoh and Laemmli model, there are more SARs per unit length of DNA in G bands than in R bands. G bands have smaller loops and a tighter 'queue' of SARs along the scaffold, so that there are more SARs per unit length of chromosome, leading to stronger staining with AT-selective stains like Giemsa.

Chromosome abnormalities

Chromosome abnormalities might be defined as changes resulting in a visible alteration of the chromosomes. How much can be seen depends on the technique used. The smallest loss or gain of material visible by traditional methods on standard cytogenetic preparations is about 4 megabases of DNA. However, fluorescence *in situ* hybridization allows much smaller changes to be seen; the development of molecular cytogenetics has removed any clear dividing line between changes described as chromosomal abnormalities and changes thought of as molecular or DNA defects. An alternative definition of a chromosomal abnormality is an abnormality produced by specifically chromosomal mechanisms. Most chromosomal aberrations are produced by misrepair of broken chromosomes, by improper recombination or by malsegregation of chromosomes during mitosis or meiosis.

Types of chromosomal abnormality

A chromosomal abnormality may be present in all cells of the body (constitutional abnormality), or may be present in only certain cells or tissues (somatic or acquired abnormality). Constitutional abnormalities must have been present very early in development, most likely the result of an abnormal sperm or egg, or maybe abnormal fertilization or an abnormal event in the early embryo. By contrast, an individual with a somatic abnormality is a mosaic, containing cells with two different chromosome constitutions, with both cell types deriving from the same zygote. Chromosomal abnormalities, whether constitutional or somatic, mostly fall into two categories: numerical and structural abnormalities. Occasionally, abnormalities have been identified in which chromosomes have the correct number and structure, but represent unequal contributions from the two parents. Perhaps unexpectedly, correct parental origin matters.

Numerical chromosomal abnormalities involve gain or loss of complete chromosomes

Three classes of numerical chromosomal abnormalities can be distinguished: polyploidy, aneuploidy and mixoploidy.

Polyploidy

Between 1 and 3% of recognized human pregnancies are triploid. The most usual cause is two sperm fertilizing a single egg (dispermy); sometimes the cause is a diploid gamete. Triploids very seldom survive to term, and the condition is not compatible with life. Tetraploidy is much rarer and always lethal. It is usually due to failure to complete the first zygotic division: the DNA has replicated to give a content of 4*C*, but cell division has not then taken place as normal. Although constitutional polyploidy is rare and lethal, all normal people have some polyploid cells.

Aneuploidy

Euploidy means having complete chromosome sets (*n*, 2*n*, 3*n*, etc.). Aneuploidy is the opposite, that is, one or more individual chromosomes extra or missing from a euploid set. Trisomy means having three copies of a particular chromosome in an otherwise diploid cell, for example trisomy 21 (47,XX or XY, +21) in Down syndrome. Monosomy is the corresponding lack of a chromosome, for example monosomy X (45,X) in Turner syndrome. Cancer cells often show extreme aneuploidy, with multiple chromosomal abnormalities. Aneuploid cells arise through two main mechanisms:

- Nondisjunction: failure of paired chromosomes to separate (disjoin) in anaphase of meiosis I, or failure of sister chromatids to disjoin at either meiosis II or at mitosis. Nondisjunction in meiosis produces gametes with 22 or 24 chromosomes, which after fertilization by a normal gamete make a trisomic or monosomic zygote. Nondisjunction in mitosis produces a mosaic.
- Anaphase lag: failure of a chromosome or chromatid to be incorporated into one of the daughter nuclei following cell division, as a result of delayed movement (lagging) during anaphase. Chromosomes that do not enter a daughter cell nucleus are lost.

Mixoploidy

Mixoploidy includes mosaicism (an individual possesses two or more genetically different cell lines all derived from a single zygote) and chimerism (an individual has two or more genetically different cell lines originating from different zygotes. Abnormalities that would be lethal in constitutional form may be compatible with life in mosaics.

Aneuploidy mosaics are common. For example, mosaicism resulting in a proportion of normal cells and a proportion of aneuploid (e.g. trisomic) cells can be ascribed to nondisjunction or chromosome lag occurring in one of the mitotic divisions of the early embryo (any monosomic cells that are formed usually die out). Polyploidy mosaics (e.g. human diploid/triploid mosaics) are occasionally found. As gain or loss of a haploid set of chromosomes by mitotic nondisjunction is most unlikely, human diploid/triploid mosaics most probably arise by fusion of the second polar body with one of the cleavage nuclei of a normal diploid zygote.

Clinical consequences of numerical abnormalities

Having the wrong number of chromosomes has serious, usually lethal, consequences. Even though the extra chromosome 21 in a man with Down syndrome is a perfectly normal chromosome, inherited from a normal parent, its presence causes multiple congenital abnormalities. Autosomal monosomies have even more catastrophic consequences than trisomies. These abnormalities must be the consequence of an imbalance in the levels of gene products encoded on different chromosomes. Normal development and function depend on innumerable interactions between gene products, including many that are encoded on different chromosomes. Altering the relative numbers of chromosomes will affect these interactions.

Having the wrong number of sex chromosomes has far fewer ill effects than having the wrong number of any autosome. 47,XXX and 47,XYY people often function within the normal range; 47,XXY men have relatively minor problems compared to people with any autosomal trisomy, and even monosomy, in 45,X women, has remarkably few major consequences.

In fact, since normal people can have either one or two X chromosomes, and either no or one Y, there must be special mechanisms that allow normal function with variable numbers of sex chromosomes. In the case of the Y chromosome, this is because it carries very few genes, whose only important function is to determine male sex. For the X chromosome, the special mechanism of lyonization controls the level of X-encoded gene products independently of the number of X chromosomes present in the cell.

Autosomal monosomy is invariably lethal at the earliest stage of embryonic life. On every chromosome there are probably a few genes where a halving of the level of the gene product is incompatible with development. Also, while such a halving is not obviously pathogenic for most genes, it may have minor effects, and the combination of hundreds or thousands of these minor effects could be enough to disrupt normal development of the embryo. Trisomies make a smaller change than monosomies in relative levels of gene products, and their effects are somewhat less. Trisomic embryos survive longer than monosomic ones, and trisomies 13, 18 and 21 are compatible with survival until birth. Interestingly, these three chromosomes seem to be relatively poor in genes. It is not so obvious why triploidy is lethal in humans and other animals. With three copies of every autosome, the dosage of autosomal genes is balanced and should not cause problems. Triploids are always sterile because triplets of chromosomes cannot pair and segregate correctly in meiosis, but many triploid plants are in all other respects healthy and vigorous. The lethality in animals is probably explained by imbalance between products encoded on the X chromosome and autosomes, which lyonization is unable to compensate.

Structural chromosomal abnormalities result from misrepair of chromosome breaks or from malfunction of the recombination system

Chromosome breaks occur either as a result of damage to DNA (by radiation or chemicals, for example) or as part of the mechanism of recombination. In G_2 phase of the cell cycle chromosomes consist of two chromatids. Breaks occurring at this stage are manifest as chromatid breaks, affecting only one of the two sister chromatids. Breaks occurring in G_1 phase, if not repaired before S phase, appear later as chromosome breaks, affecting both chromatids. Cells have enzyme systems that recognize and if possible repair broken chromosome ends. Repair can be either by joining two broken ends together, or by capping a broken end with a telomere. Cell cycle checkpoint mechanisms normally prevent cells with unrepaired chromosome breaks from entering mitosis; if the damage cannot be repaired, the cell commits suicide (apoptosis). Structural abnormalities arise when breaks are repaired incorrectly. Provided there are no free broken ends, the cell cycle checkpoints can be negotiated satisfactorily, but in fact the wrong broken ends may have been joined together. Any resulting chromosome that has no centromere (acentric) or two centromeres (dicentric) will not segregate stably in mitosis, and will eventually be lost. Chromosomes with a single centromere can be stably propagated through successive rounds of mitosis, even if they are structurally abnormal. Meiotic recombination between mispaired chromosomes is a common cause of translocations, especially in spermatogenesis.

Structural chromosomal abnormalities are *balanced* if there is no net gain or loss of chromosomal material, and *unbalanced* if there is net gain or loss. In general, balanced abnormalities (inversions, balanced translocations) have no effect on the phenotype, although there are important exceptions to this:

- a chromosome break may disrupt an important gene;
- the break may affect expression of a gene even though it does not disrupt the coding sequence. It may separate a gene from a control element, or it may put the gene in an inappropriate chromatin environment, for example translocating a normally active gene into heterochromatin;
- balanced X-autosome translocations cause problems with X-inactivation.

Robertsonian translocations are sometimes called centric fusions, but this is misleading because in fact the breaks are in the proximal short arms. The translocation chromosome is really dicentric, but because the two centromeres are very close together they function as one, and the chromosome segregates regularly. The distal parts of the two short arms are lost as an acentric fragment. Short arms of acrocentric chromosomes contain only arrays of repeated ribosomal RNA genes, and the loss of two short arms has no phenotypic effect. Because there is no phenotypic effect, Robertsonian translocations are regarded as balanced, even though in fact some material has been lost.

Unbalanced abnormalities can arise directly, through deletion or, rarely, duplication, or indirectly by malsegregation of chromosomes during meiosis in a carrier of a balanced abnormality. Carriers of balanced structural abnormalities can run into trouble during meiosis, if the structures of homologous pairs of chromosomes do not correspond:

- A carrier of a balanced reciprocal translocation can produce gametes that after fertilization give rise to an entirely normal child, a phenotypically normal balanced carrier, or various unbalanced karyotypes that always combine monosomy for part of one of the chromosomes with trisomy for part of the other. It is not possible to make general statements about the relative frequencies of these outcomes. The size of any unbalanced segments depends on the position of the breakpoints. If the unbalanced segments are large, the fetus will probably abort spontaneously; unbalance for smaller segments may result in liveborn abnormal babies.
- A carrier of a balanced Robertsonian translocation can produce gametes that after fertilization give rise to an entirely normal child, a phenotypically normal balanced carrier, or a conceptus with full trisomy or full monosomy for one of the chromosomes involved.
- A carrier of a pericentric inversion may produce unbalanced offspring because when the inverted and noninverted homologs pair they form a loop so that matching segments pair along the whole length of the chromosomes. If a crossover occurs within the loop, the result is a chromosome carying an unbalanced deletion and duplication. Paracentric chromosome inversions form similar loops, but any crossover within the loop generates an acentric or dicentric chromosome, which is unlikely to survive.

Apparently normal chromosomal complements may be pathogenic if they have the wrong parental origin

The rare abnormalities described below demonstrate that it is not enough to have the correct number and structure of chromosomes; they must also have the correct parental origin.

46,XX conceptuses in which both genomes originate from the same parent (uniparental diploidy) never develop correctly. For some individual chromosomes, having both homologs derived from the same parent (uniparental disomy) also causes abnormality. A small number of genes are imprinted with their parental origin and are expressed differently according to the origin. It is assumed that the abnormalities of uniparental disomy and uniparental diploidy are caused by abnormal expression of such imprinted genes.

Uniparental diploidy is seen in hydatidiform moles, abnormal conceptuses with a 46,XX karyotype of exclusively paternal origin. Molar pregnancies show widespread hyperplasia of the trophoblast but no fetal parts, and have a significant risk of transformation into choriocarcinoma. Genetic marker studies show that most moles are homozygous at all loci, indicating that they arose by chromosome doubling from a single sperm. Ovarian teratomas are the result of maternal uniparental diploidy. These rare benign tumors of the ovary consist of disorganized embryonic tissues, without extraembryonic membranes. They arise by activation of an unovulated oocyte.

Uniparental disomy (UPD), affecting a single pair of homologs, goes undiagnosed if the result is not abnormal, but is detected for chromosomes for which it produces characteristic syndromes. UPD can be isodisomy, where both homologs are identical, or heterodisomy, where they are derived from both homologs in one parent. The usual cause is thought to be trisomy rescue: a conceptus that is trisomic and would otherwise die, occasionally loses one chromosome by mitotic nondisjunction or anaphase lag from a totipotent cell. The euploid progeny of this cell form the embryo, while all the aneuploid cells die. If each of the three copies has an equal chance of being lost, there will be a two in three chance of a single chromosome loss leading to the normal chromosome constitution and a one in three chance of uniparental disomy (either paternal or maternal). Uniparental isodisomy may possibly arise by selection pressure on a monosomic embryo to achieve euploidy by selective duplication of the monosomic chromosome.

3

GENES IN PEDIGREES

MENDELIAN PEDIGREE PATTERNS

The simplest genetic characters are those whose presence or absence depends on the genotype at a single locus. That is not to say that the character itself is programmed by only one pair of genes - expression of any human character is likely to require a large number of genes and environmental factors. However, sometimes a particular genotype at one locus is both necessary and sufficient for the character to be expressed, given the normal genetic and environmental background of the organism. Such characters are called mendelian. In humans over 10 000 mendelian characters are known. As described in the *Introduction*, the essential starting point for acquiring information on any human mendelian character, whether pathological or non-pathological, is the OMIM Internet database.

Dominance and recessiveness are properties of characters, not genes

A character is dominant if it is manifest in the heterozygote and recessive if not. Note that dominance and recessiveness are properties of characters, not genes. Thus sickle cell anemia is recessive because only HbS homozygotes manifest it. Heterozygotes for the same gene show sickling trait, which is therefore a dominant character. Most human dominant syndromes are known only in heterozygotes. Sometimes homozygotes have been described, born from matings of two heterozygous affected people, and often the homozygotes are much more severely affected. Examples are achondroplasia (short-limbed dwarfism) and Type 1 Waardenburg syndrome (deafness with pigmentary abnormalities). Nevertheless we describe achondroplasia and Waardenburg syndrome as dominant because these terms describe phenotypes seen in heterozygotes. In experimental organisms, where this uncertainty does not exist, geneticists tend to use the term semidominant when the heterozygote has an intermediate phenotype, reserving 'dominant' for conditions where the homozygote is indistinguishable from the heterozygote - Huntington disease (adult-onset progressive neurological deterioration) for example. The question of dominance has been well reviewed by Wilkie (1994). Males are hemizygous for loci on the X and Y chromosomes, where they have only a single copy of each gene, so the question of dominance or recessiveness does not arise in males for X- or Y-linked characters.

There are five basic mendelian pedigree patterns

A female heterozygous for an X-linked condition (dominant or recessive), is a mosaic. Each cell expresses either the normal or the abnormal allele, but not both. Where the phenotype depends on a circulating product, as in hemophilia (failure of blood to clot), there is an averaging effect between the normal and abnormal cells. Female carriers have an intermediate phenotype, and are usually clinically unaffected but biochemically abnormal. Where the phenotype is a localized property of individual cells, as in hypohidrotic ectodermal dysplasia (MIM 305100: missing sweat glands, abnormal teeth and hair) female carriers show patches of normal and abnormal tissue. Occasional manifesting heterozygotes are seen for X-linked recessive conditions: these women may be quite severely affected because by bad luck most cells in some critical tissue have inactivated the normal X.

There are probably no Y-linked diseases

Although a few Y-linked characters have been described, no Y-linked diseases are known, apart from disorders of male sexual function. Conceivably such a disease may exist undiscovered, but this is unlikely for two reasons. First, the pedigree pattern would be strikingly noticeable, especially in societies that trace family through the male line, yet they have not been noted (claims for 'porcupine men' are dubious, see MIM 146600). Second, the Y-chromosome cannot carry any genes whose function is important for health, because females are perfectly normal without any Y-linked genes. Thus any Y-linked genes must code either for non-essential characters or for male-specific functions, and defects are unlikely to cause diseases apart from defects of male sexual function. Genes present as functional copies on both the Y and the X might prove an exception to this argument.

The mode of inheritance can rarely be defined unambiguously in a single pedigree

Given the limited size of human families, it is rarely possible to be completely certain of the mode of inheritance of a character simply by inspecting a single pedigree. In experimental animals one would set up a test cross and check for a 1 in 2 or 1 in 4 ratio. In human pedigrees the proportion of affected children is not a very reliable indicator. Mostly this is because the numbers are too small, but in addition, the way in which the family was ascertained can bias the ratio of affected to unaffected children observed. For recessive conditions, the proportion of affected children often seems to be greater than 1 in 4. This is because families are normally ascertained when they have an affected child; families where both parents are carriers but, by good fortune, nobody is affected are systematically missed.

For many of the rarer conditions, the stated mode of inheritance is no more than an informed guess. Assigning modes of inheritance is important, because that is the basis of the risk estimates used in genetic counseling. However, it is important to recognize that the modes of inheritance are often working hypotheses rather than established fact. OMIM uses an asterisk to denote entries with relatively well-established modes of inheritance. Only when a cloned copy of the gene is available can the inheritance be determined with certainty.

One gene - one enzyme does not imply one gene - one syndrome

Pedigree patterns provide the essential entry point into human genetics, but they are only a starting point for defining genes. It would be a serious error to imagine that the 10 000 or so known mendelian characters define 10 000 DNA coding sequences. This would be an unjustified extension of the one gene - one enzyme hypothesis of Beadle and Tatum. Back in the 1940s this hypothesis allowed a major leap forward in understanding how genes determine phenotypes. Since then it has been extended - some genes encode nontranslated RNAs, some proteins are not enzymes, and many proteins contain several separately encoded polypeptide chains. But even with these extensions, Beadle and Tatum's hypothesis cannot be used to imply a one-to-one correspondence between entries in the OMIM catalogue and entries in Genbank, the DNA sequence database.

The genes of classical genetics are abstract entities. Any character that is determined at a single chromosomal location will segregate in a mendelian pattern - but the determinant may not be a gene in the molecular geneticist's sense of the word. Fascio-scapulo-humeral muscular dystrophy (MIM 158900: severe but nonlethal weak ness of certain muscle groups) is associated with small deletions of sequences at 4q35, but nobody (at the time of writing) has managed to find a protein-coding sequence at that location, despite intensive searching and sequencing. The 'gene' for Charcot-Marie-Tooth disease type 1A (MIM 118220: motor and sensory neuropathy) turned out to be a 1.5 Mb tandem duplication on chromosome 17p11.2. These examples are unusual; most OMIM entries probably do describe the consequences of mutations affecting a single transcription unit, but because of locus and allelic heterogeneity, there is still no one-to-one correspondence with Genbank entries.

Locus heterogeneity is common in syndromes that result from failure of a complex pathway: Profound congenital hearing loss is often genetic, and when genetic it is usually autosomal recessive. However, when two people with autosomal recessive profound hearing loss marry, as they often do, the children usually have normal hearing. This is an example of complementation. The children will have normal hearing whenever the parents carry mutations in different genes. Diseases and developmental defects represent the failure of a pathway. It is easy to see that many different genes would be needed to construct so exquisite a machine as the cochlear hair cell, and a defect in any of those genes could lead to deafness. Such locus heterogeneity is only to be expected in conditions like deafness, blindness or mental retardation, where a rather general pathway has failed; but even with more specific pathologies, multiple loci are very frequent. A striking example is Usher syndrome, an autosomal recessive combination of hearing loss and retinitis pigmentosa, which can be caused by mutations at eight or more unlinked loci (Smith *et al.*, 1994). OMIM has separate entries for known examples of locus heterogeneity (defined by linkage or mutation analysis), but there must be many undetected examples still contained within single entries.

Allelic series are a cause of clinical heterogeneity: Sometimes several apparently distinct human phenotypes turn out to be all caused by different allelic mutations at the same locus. The difference may be one of degree - mutations that partially inactivate the dystrophin gene produce Becker muscular dystrophy, while mutations that completely inactivate the same gene produce the similar but more severe Duchenne muscular dystrophy (lethal muscle wasting). Sometimes the difference is qualitative - inactivation of the androgen receptor gene causes androgen insensitivity (MIM 313700; 46,XY embryos develop as females), but expansion of a run of glutamine codons within the same gene causes a very different disease, spinobulbar muscular atrophy or Kennedy disease (MIM 313200).

Mitochondrial inheritance gives a recognizable matrilinear pedigree pattern

In addition to the mutations in genes carried on the nuclear chromosomes, mitochondrial mutations are a significant cause of human genetic disease. The mitochondrial genome (see *Figure 7.2*) is small but highly mutable compared to nuclear DNA, probably because mitochondrial DNA replication is more error-prone and the number of replications is much higher. Mitochondrially-encoded diseases have two unusual features, matrilineal inheritance and frequent heteroplasmy (Wallace, 1994).

Inheritance is matrilineal, because sperm do not contribute mitochondria to the zygote (this assertion rests on limited evidence; however, paternally-derived mitochondrial variants are not detected in children). Thus a mitochondrially inherited condition can affect both sexes, but is passed on only by affected mothers, giving a recognizable pedigree pattern.

Cells contain many mitochondrial genomes. In some patients with a mitochondrial disease, every mitochondrial genome carries the causative mutation (homoplasmy), but in other cases a mixed population of normal and mutant genomes is seen within each cell (heteroplasmy). Unlike nuclear genetic mosaicism, which must arise post-zygotically, mitochondrial heteroplasmy can be transmitted from heteroplasmic mother to heteroplasmic child. In such cases the proportion of abnormal mitochondrial genomes can vary remarkably between the mother and child, suggesting that a surprisingly small number of maternal mitochondrial DNA molecules give rise to all the mitochondrial DNA of the child.

One of the best known mitochondrial diseases shows an unusual mode of inheritance: Leber's hereditary optic atrophy (LHON, MIM 535000: sudden and irreversible loss of sight) is associated with various mitochondrial mutations and is inherited matrilinearly, but unexpectedly, almost all affected patients are male. Possibly LHON requires both a mitochondrial and an X-linked mutation, but attempts to demonstrate an X-linked susceptibility have not been successful, so the reason for the male excess remains unknown.

CELL-BASED DNA CLONING

Fundamentals of DNA technology and the importance of DNA cloning

The fundamentals of current DNA technology are very largely based on two quite different approaches to studying specific DNA sequences within a complex DNA population:

- DNA cloning. The desired fragment must be *selectively amplified* so that it is purified essentially to homogeneity. Thereafter, its structure and function can be comprehensively studied, for example by DNA sequencing, *in vitro* expression studies, etc., and various manipulations can be achieved to change its structure by *in vitro* mutagenesis.

- Molecular hybridization. The fragment of interest is not amplified, but instead is *specifically detected* within a complex mixture of many different sequences. Its chromosomal location can be determined in this way and some information can be gained regarding its structure. If expressed, the sequence of interest can be detected within a complex RNA or cDNA population from specific cells, enabling comprehensive analysis of its expression patterns.

Before DNA cloning, our knowledge of DNA was extremely limited. DNA cloning technology changed all that and revolutionized the study of genetics. Why was DNA cloning such an important technological advance? One important consideration is the tremendous size and complexity of DNA sequences (compared to, say, protein sequences). Individual nuclear DNA molecules contain hundreds of millions of nucleotides. When DNA is isolated from cells using standard methods, these huge molecules are fragmented by shear forces, generating complex mixtures of still very large DNA fragments (typically 50–100 kb long).

Given the above, how can relatively homogeneous DNA populations be prepared from such a complex starting mixture. In the case of DNA from human cells and a wide variety of complex eukaryotic cells, one early approach had been to separate different classes of DNA fragments according to their base composition. The DNA preparation was submitted to ultracentrifugation in equilibrium density gradients (e.g. in CsCl density gradients). When this was achieved, the DNA was fractionated into a major band (the bulk DNA) and several minor satellite DNA bands. The satellite DNA species have different buoyant densities to the bulk DNA because they have unusual sequences and their base composition is different to the majority of the DNA in cells (these properties in turn reflect the involvement of satellite DNA in specific aspects of chromosome structure and function. Although valuable and interesting, the purified satellite DNAs were, however, a minor component of the genome and did not contain genes.

What DNA cloning offered was a general method for studying *any* DNA sequence.

DNA cloning is a general method of selectively amplifying DNA sequences to generate homogenous DNA populations

In order to study genes, methods had to be developed to purify them. Because mammalian genomes are complex, any specific gene or DNA fragment of interest normally represents only a tiny fraction of the total DNA in a cell. For example, the β-globin gene comprises only 0.00005% of the 3300 megabases (Mb) of human genomic DNA, and even the massive 2.5 Mb dystrophin gene, the largest gene that has been identified, accounts for only about 0.08% of human genomic DNA.

One way of enriching for gene sequences is to isolate total RNA, or poly(A)$^+$ messenger RNA (mRNA) from suitable cells and convert this to complementary DNA (cDNA) using the enzyme reverse transcriptase. In some cases, this can result in a profound enrichment for specific exonic DNA sequences when the relevant genes are known to be expressed at very high levels in a specific cell type. In most cases, however, the desired gene sequences still represent only a tiny proportion of the total cDNA population.

In order to have a general method of studying a specific DNA sequence within a complex DNA population, the technology of DNA cloning was developed. The essential characteristic of DNA cloning is that the desired DNA fragments must be *selectively amplified* in some way, resulting in a programmed large increase in copy number of selected DNA sequences. In practice, this involves multiple rounds of DNA replication catalyzed by a DNA polymerase acting on one or more types of template DNA molecule. Essentially two different DNA cloning approaches are used:

- Cell-based DNA cloning. This was the first form of DNA cloning to be developed, and is an *in vivo* cloning method. The first step in this approach involves attaching foreign DNA fragments *in vitro* to DNA sequences which are capable of independent replication. The recombinant DNA fragments are then transferred into suitable host cells where they can be propagated selectively. In the past, the term DNA cloning has been used exclusively to signify this particular approach.

- Cell-free DNA cloning. The polymerase chain reaction (PCR) is a newer form of DNA cloning which is enzyme mediated and is conducted entirely *in vitro*.

Principles of cell-based DNA cloning

Cell-based DNA cloning requires attachment in vitro of DNA fragments to purified replicons and propagation in suitable host cells

The essence of cell-based DNA cloning involves four steps:

i. *Construction of recombinant DNA molecules* by *in vitro* covalent attachment (ligation) of the desired DNA fragments (target DNA) to a replicon (any sequence capable of independent DNA replication). This step is facilitated by cutting the target DNA and replicon molecules with specific restriction endonucleases before joining the different DNA fragments using the enzyme DNA ligase.

ii. *Transformation.* The recombinant DNA molecules are transferred into host cells (often bacterial or yeast cells) in which the chosen replicon can undergo DNA replication independently of the host cell chromosome(s).

iii. *Selective propagation of cell clones* involves two stages. Initially the transformed cells are plated out by spreading on an agar surface in order to encourage the growth of *well-separated* cell colonies. These are cell clones (populations of identical cells all descended from a single cell). Subsequently, *individual* colonies can be picked from a plate and the cells can be further expanded in liquid culture.

iv. *Isolation of recombinant DNA clones* by harvesting expanded cell cultures and selectively isolating the recombinant DNA.

Replicons and host cells: Cell cloning requires that the foreign DNA fragments which are introduced into a host cell must be able to replicate. If not, the foreign DNA would soon be diluted out as the host cell undergoes many rounds of cell division. However, foreign DNA fragments will generally lack an origin of replication that will function in the host cell. They require, therefore, to be attached to an independent replicon so that their replication is controlled by the replicon's origin of replication.

In principle, the necessary replicon could be provided in two ways. One possibility is to introduce the fragments into the host cells in such a way that in each cell one, or a very few, fragments integrate into a host cell chromosome and are propagated under the control of a host cell replicon. This is the way that retroviruses, for example, integrate their DNA into host cell chromosomes.

However, as a general method for cloning, this approach suffers from numerous disadvantages, including the difficulty in retrieving the inserted DNA. Instead, cell-based cloning very largely relies on a different approach: the foreign DNA fragments are attached *in vitro* to a purified replicon and the resulting hybrid molecules are transferred into host cells, where they replicate *independently of the host cell chromosomes.* Because the foreign DNA fragments (target DNA) can be viewed as passengers of the replicon, replicons used for cloning are described as vector molecules. Notice that, although the resulting replication of the introduced DNA fragments is independent of the host chromosomes, the vector may have an origin of replication that originates from either a natural extrachromosomal replicon or, in some cases, a chromosomal replicon.

Although some DNA cloning systems involve human and other mammalian cells as hosts, the great bulk of cell-based DNA cloning has used modified bacterial or fungal host cells. The former cells are widely used because of their capacity for rapid cell division. Bacterial cell hosts have a single circular double-stranded chromosome with a single origin of replication.

Replication of the host chromosome subsequently triggers cell division so that each of the two resulting daughter cells contains a single chromosome like their parent cell (i.e. the copy number is maintained at one copy per cell). However, the replication of extrachromosomal replicons is not constrained in this way: many such replicons go through several cycles of replication during the cell cycle and can reach high copy numbers.

Extrachromosomal replicons as vector molecules: Two basic types of extrachromosomal replicons are found in bacterial cells:

- Plasmids are small circular double-stranded DNA molecules which individually contain very few genes. Their existence is intracellular, being vertically distributed to daughter cells following host cell division, but they can be transferred horizontally to neighboring cells during bacterial conjugation. Natural examples include plasmids which carry the sex factor (F) and those which carry drug-resistance genes.
- Bacteriophages are viruses which infect bacterial cells. DNA-containing bacteriophages often have genomes containing double-stranded DNA which may be circular or linear. Unlike plasmids, they can exist extracellularly. The mature virus particle (virion) has its genome encased in a protein coat so as to facilitate adsorption and entry into a new host cell.

In order for naturally occurring replicons to be used as vector molecules for cell-based DNA cloning, various modifications need to be made. Similarly, the host cells that are used for cloning are specialized cells whose genotype has been selected to optimize their use in DNA cloning. Typically, cloning systems are designed to ensure that joining of the foreign DNA fragment occurs at a unique location in the vector molecule. Additionally, they have in-built selection systems so that cells which contain the relevant vector molecule can be specifically selected. In many cases, there are additional screening systems to ensure detection and propagation of cells containing recombinant DNA.

Restriction endonucleases enable the target DNA to be cut into pieces of manageable size and facilitate ligation to similarly cut vector molecules

A major boost to the development of cell-based DNA cloning was the discovery and exploitation of type II restriction endonucleases, enzymes which normally cleave DNA whenever a small, specific recognition sequence, usually 4–8 base pairs (bp) long occurs.

The recognition sequences for the vast majority of type II restriction endonucleases are normally palindromes, that is the sequence of bases is the same on both strands when read in the 5' → 3' direction, as a result of a twofold axis of symmetry. In some cases, the cleavage points occur exactly on the axis of symmetry, giving products (restriction fragments) which are blunt-ended. In most cases, however, the cleavage points do not fall on the symmetry axis, so that the resulting restriction fragments possess so-called 5' overhangs or 3' overhangs.

Overhanging ends generated by cleavage with a restriction nuclease are often described as sticky ends or cohesive termini because the two overhanging ends of each fragment are complementary in base sequence, and will have a tendency to associate with each other, or with any other similarly complementary overhang, by forming base pairs. Different fragments with the same sequences in their overhanging ends can be generated by: (i) cutting with the same restriction nuclease; (ii) cutting with different restriction endonucleases that happen to recognize the same target sequence (isoschizomers); or (iii) by cutting with enzymes which have different recognition sequences but happen to produce compatible sticky ends, for example *Bam*HI and *Mbo*I.

The termini of restriction fragments which have the same type of overhanging ends can associate in a variety of different ways, either intramolecularly (cyclization), or between molecules to form linear concatemers or circular compound molecules. Intermolecular reactions occur most readily at high DNA concentrations. At very low DNA concentrations, however, individual termini on different molecules have less opportunity of making contact with each other, and intramolecular cyclization is favored.

Because the overhanging ends generated by restriction endonucleases are very short (typically four nucleotides or less), hydrogen bonding between complementary overhanging ends provides a rather weak contact between two molecules, and can only be maintained at low temperatures. However, it does facilitate subsequent covalent bonding between the two associated molecules (DNA ligation). This is performed using the enzyme DNA ligase. Ligation of blunt-ended fragments is also possible, although less efficient than sticky end ligation.

Generally, ligation reactions are designed to promote the formation of recombinant DNA (by ligating target DNA to vector DNA), although vector cyclization, vector-vector concatemers and target DNA-target DNA ligation are also possible. To achieve this, the vector molecules are often treated so as to prevent or minimize their ability to undergo cyclization. There are two common ways of achieving this.

- *Cutting of vectors with two different restriction endonucleases.* Often, vector molecules have multiple unique restriction sites, in which foreign DNA can be cloned, occurring in a short segment of the molecule. It is often convenient, therefore, to cut the vector with two restriction endonucleases which do not produce complementary overhanging ends (e.g. *Eco*RI and *Bam*HI), and remove the small vector fragment between the sites, resulting in a vector molecule whose two ends cannot religate. However, if target DNA is cut with the same enzyme combination, recombinant DNA can easily be formed.
- *Vector dephosphorylation.* During DNA ligation *in vitro*, the enzyme DNA ligase will catalyze the formation of a phosphodiester bond only if one nucleotide contains a 5' phosphate group and the other contains a 3' hydroxyl group. If the 5' phosphate groups at both ends of the vector DNA are removed by treatment with alkaline phosphatase, the tendency for the vector DNA to recircularize will therefore be minimized. A foreign DNA

insert can, however, provide 5' -terminal phosphates which can then be joined to the 3' hydroxyl groups provided by the vector. This method, therefore, increases the frequency of cells containing recombinant DNA.

Introducing recombinant DNA into recipient cells provides a method for fractioncting a complex starting DNA population

The plasma membrane of cells is selectively permeable and does not normally admit large molecules such as long DNA fragments. However, cells can be treated in certain ways (e.g. by exposure to certain high ionic strength salts, short electric shocks, etc.) so that the permeability properties of the plasma membranes are altered. As a result, a fraction of the cells become competent, that is capable of taking up foreign DNA from the extracellular environment. Only a small percentage of the cells will take up the foreign DNA (DNA transformation). However, those that do will often take up only a single molecule (which can, however, subsequently replicate many times within a cell). *This is the basis of the critical fractionation step in cell-based DNA cloning*; the population of transformed cells can be thought of as a sorting office in which the complex mixture of DNA fragments is sorted by depositing individual DNA molecules into individual recipient cells.

Because circular DNA (even nicked circular DNA) transforms much more efficiently than linear DNA, most of the cell transformants will contain cyclized products rather than linear recombinant DNA concatemers and, if an effort has been made to suppress vector cyclization (e.g. by dephosphorylation), most of the transformants will contain recombinant DNA. Note, however, that cotransformation events (the occurrence of more than one type of introduced DNA molecule within a cell clone,) may be comparatively common in some cloning systems.

The transformed cells are allowed to multiply. In the case of cloning using plasmid vectors and a bacterial cell host, a solution containing the transformed cells is simply spread over the surface of nutrient agar in a petri dish (plating out). This usually results in the formation of bacterial colonies which consist of cell clones (identical progeny of a single ancestral cell). Picking an individual colony into a tube for subsequent growth in liquid culture permits a secondary expansion in the number of cells which can be scaled up to provide very large yields of cell clones, all identical to an ancestral single cell. If the original cell contained a single type of foreign DNA fragment attached to a replicon, then so will the descendants, resulting in a huge amplification in the amount of the specific foreign fragment.

Expanded cultures representing cell clones derived from a single cell can then be processed to recover the recombinant DNA. To do this, the cells are lysed, and the DNA is extracted and purified using procedures that result in recovery of the recombinant DNA. Such procedures take advantage of differences between the recombinant DNA and the host chromosomal DNA. In the case of bacterial cells, the bacterial chromosome is circular double-stranded DNA, like any plasmids containing introduced foreign DNA. However, the chromosomal DNA is relatively very large (~4.6 Mb) compared with most recombinant DNA molecules (often only a few kilobases long). During cell lysis and subsequent extraction procedures, the very large bacterial chromosomal DNA, but not the small recombinant plasmid DNA, will undergo nicking and shearing, generating linear DNA fragments with free ends. This difference can be exploited by subjecting the isolated DNA to a denaturation step, often as a result of exposure to alkaline pH in the 12.0–12.5 range.

Following this treatment, the linearized host cell DNA readily denatures, but the strands of covalently closed circular (CCC) plasmid DNA are unable to separate. After normal conditions have been restored, the two strands of the CCC DNA rapidly re-align in perfect register to form native superhelical molecules or so-called supercoiled DNA (this higher order form of twisting occurs in any CCC DNA because the tension introduced by twisting of the double helix cannot be relaxed, unlike in DNA with one or two free ends where relaxation is possible by rotation of a free end). The denatured host cell DNA precipitates out of solution, leaving behind the CCC plasmid DNA.

If required, further purification is possible by column chromatography. Alternatively, equilibrium density gradient centrifugation (isopycnic centrifugation) is used: the partially purified DNA is centrifuged to equilibrium in a solution of cesium chloride containing ethidium bromide (EtBr). EtBr binds by *intercalating* between the base pairs, thereby causing the DNA helix to unwind. Unlike chromosomal DNA, a CCC plasmid DNA has no free ends and can only unwind to a limited extent, which limits the amount of EtBr it can bind. EtBr-DNA complexes are denser when they contain less EtBr, so CCC plasmid DNA will band at a lower position in the cesium chloride gradient than chromosomal DNA and plasmid circles that are open, enabling separation of the recombinant DNA from host cell DNA. The resulting recombinant DNA molecules will normally be identical to each other (representing a single target DNA fragment) and are referred to as DNA clones.

DNA libraries are a comprehensive set of DNA clones representing a complex starting DNA population

The first attempts at cloning human DNA fragments in bacterial cells concentrated on target sequences which were highly abundant in a particular starting DNA population. For example, most human cells contain much the same complex collection of DNA sequences in the nucleus, but the mRNA populations of different cell types can be quite different. Although the mRNA population in each cell is complex, some cells are particularly devoted to synthesizing a specific type of protein and so their mRNA populations have a few predominant mRNA species (e.g. much of the mRNA made in erythrocytes is α and β globin mRNA). The enzyme reverse transcriptase (RNA-dependent DNA polymerase) can be used to make a DNA copy that is complementary in base sequence to the mRNA, so-called cDNA. Hence, cDNA from erythrocytes is greatly enriched in globin cDNA, facilitating its isolation.

Modern DNA cloning approaches offer the possibility of making comprehensive collections of DNA clones (DNA libraries) from extremely complex starting DNA populations (such as total human genomic DNA). This approach enables DNA sequences which are very rare in the starting population to be represented in a library of DNA clones, whence they can be isolated individually by selecting a suitable host cell colony and amplifying it. Two basic varieties of this method have popularly been undertaken: the construction of genomic DNA libraries and cDNA libraries.

Newly constructed libraries are said to be *unamplified*, although this is a misleading term because the initially transformed cells have been amplified to form separated cell colonies. Often, cell colony formation is allowed to proceed on top of membranes that are overlaid on to the surface of nutrient agar in sterile culture dishes. Copies of the library can then be made by replica plating on to a similar sized membrane prior to overlaying on to a nutrient agar surface and colony growth. More recently, individually picked cell colonies have been spotted in gridded arrays on to suitable membranes or into the wells of microtiter dishes where they can be stored for long periods at -70°C in the presence of a cell-stabilizing medium such as glycerol. For multiple distribution, amplified libraries are required. The cells from representative primary filters are washed off into cell culture medium, diluted and stabilized by the presence of glycerol, or some alternative stabilizing agent. Individual aliquots can then be plated out at a later stage to regenerate the library. This additional amplification step, however, may result in distortion of the original representation of cell clones because during the amplification stage there may be differential rates of growth of different colonies.

Genomic DNA libraries: In the case of complex eukaryotes, such as mammals, all nucleated cells have essentially the same DNA content, and it is often convenient to prepare a genomic library from easily accessible cells, such as blood cells. The starting material is genomic DNA which has been fragmented in some way, usually by digesting with a restriction endonuclease. Typically, the genomic DNA is digested with a 4 bp cutter such as *Mbo*I which recognizes the sequence GATC. This sequence will occur about every 275 bp on average in human genomic DNA, so that there will be few DNA sequences which lack a recognition site for this enzyme. Clearly, complete digestion of the starting DNA with this enzyme would produce very small fragments. Instead, the extent of enzymic cleavage is deliberately minimized: under conditions of partial restriction digestion (low enzyme concentration, short time of incubation, etc.), cleavage will occur at only a small number of the potential restriction sites. This clearly has the benefit of being able to produce large fragments for cloning. Importantly, it also allows random fragmentation of the DNA. Thus, for a specific sequence location, the pattern of cutting will be different on different copies of the same starting DNA sequence. Such random fragmentation ensures that the library will contain as much representation as possible of the starting DNA. Additionally, it has the advantage that it results in clones with overlapping inserts. As a result, after characterization of the insert of one clone, attempts can be made to access clones from the same general region by identifying those with inserts showing some similarities to that of the original clone.

The complexity (number of independent DNA clones) of a genomic DNA library can be defined in terms of genome equivalents (GE). A genome equivalent of 1, a so-called one-fold library, is obtained when the number of independent clones = genome size/average insert size. For example, for a human genomic DNA library that has an average insert size of 40 kb, 1 GE = 3000 Mb/40 kb = 75 000 independent clones. A library such as this which has 300 000 clones is sometimes termed a 4-fold library because it has 4 GE. Because of sampling variation, however, the number of GEs must be considerably greater than 1 to have a high chance of including any particular sequence within that library. Consequently, attempts are normally made to prepare libraries with several GEs.

Subgenomic DNA libraries have also been a valuable resource: the ability to fractionate individual human chromosomes or specific chromosome bands has permitted construction of chromosome-specific libraries and chromosome band-specific libraries, obtained following chromosome dissection.

cDNA libraries: As gene expression can vary in different cells and at different stages of development, the starting material for making cDNA libraries is usually total RNA from a specific tissue or specific developmental stage of embryogenesis. From this, poly(A)$^+$ mRNA can be selected by specific binding to a complementary single-stranded oligonucleotide or polynucleotide which is bound to a solid matrix. For example, chromatography columns containing poly(U) bound to sepharose or oligo(dT) bound to cellulose have been widely used: the poly(A)$^+$ mRNA selectively binds to the poly(U) or oligo(dT) components and subsequently can be eluted using buffers of high ionic strength to disrupt the hydrogen bonding. The isolated poly(A)$^+$ mRNA can then be converted, using reverse transcriptase, to a double-stranded cDNA copy. To assist cloning, oligonucleotide linkers which contain suitable restriction sites are ligated to each end of the cDNA.

Recombinant screening is often achieved by insertional inactivation of a marker gene

An essential requirement for cell-based DNA cloning systems is a method of detecting cells containing the appropriate vector molecule, and the subset which contains recombinant DNA.

Screening for cells transformed by vector molecules: Identification of cells containing the vector molecule requires engineering or selection of the vector molecule to contain a suitable marker gene whose expression provides a means of identifying cells containing it. Two popularly used marker gene systems are based on:

- antibiotic resistance genes. A host cell strain is chosen that is sensitive to a particular antibiotic, often ampicillin, tetracycline or chloramphenicol. The corresponding vector has been engineered to contain a gene which confers resistance to the antibiotic. After transformation, cells are plated on agar containing the antibiotic to rescue cells transformed by the vector.
- β-galactosidase gene complementation. The host cell is a mutant which contains a fragment of the β-galactosidase gene but cannot make any functional β-galactosidase. The vector is engineered to contain a different fragment of the β-galactosidase gene. After transformation by the vector, functional

complementation occurs resulting in active β-galactosidase which can be assayed by acting on a colorless substance, Xgal (5-bromo, 4-chloro, 3-indolyl β-d-galactopyranoside), to make a blue product.

Recombinant screening: Identification of cells containing recombinant DNA (vector molecules with inserts) is often accomplished by insertional inactivation of the marker gene. The vector molecule is designed to have a *multiple cloning site* (*polylinker*) located within the marker gene. The number of nucleotides in the polylinker sequence containing the multiple cloning sites is a multiple of three, so that its insertion in the marker gene does not result in a shift in the translation reading frame.

As a result of maintaining the reading frame, and the very small size of the polylinker, the activity of the marker gene is maintained despite the insertion of the polylinker. However, if the vector then contains recombinant DNA inserted in the middle of the polylinker, the resulting large insertion causes loss of expression of the marker gene. In the case of a marker β-galactosidase gene, insertional inactivation means that cells containing recombinant DNA are colorless in the presence of Xgal, while cells containing nonrecombinant vector are blue.

Another common screening system involves the use of suppressor tRNA genes, mutant tRNA genes which can recognize a normal termination codon signal and, in response, introduce an amino acid in the polypeptide chain. This is possible because the mutation in such tRNA genes (often mutated $tRNA^{Glu}$ or $tRNA^{Tyr}$ genes) results in an altered anticodon sequence that is complementary to one of the normal termination codons: UAA (ochre), UAG (amber) or UGA (opal).

In such systems, the host cell often carries a defective marker gene that is designed to have a premature stop codon (i.e. an amber, ochre or opal mutation) resulting in a phenotype that can be easily scored. If the vector carries a suppressor tRNA gene that suppresses the effect of this termination codon, the marker gene activity will be recovered and the wild-type phenotype will be restored. However, if the suppressor tRNA gene is inactivated by insertional inactivation, a mutant phenotype will again be produced.

If there is already available a DNA probe that is closely related to the desired recombinant DNA, colonies containing the latter can be detected directly by colony hybridization.

Initial characterization of recombinant DNA clones involves restriction mapping

Restriction mapping of DNA clones involves cutting the DNA with one or more of a series of different restriction nucleases and separating the resulting fragments according to size by agarose gel electrophoresis. Because of the conspicuous deficiency in the CpG dinucleotide in vertebrate genomes, recognition sites that are GC-rich will occur comparatively less frequently than expected in vertebrate DNA, but will not be so rare in bacterial cell DNA.

Double digests (cleavage by two different enzymes) and partial digests (reduced digestion so that not every cleavage site is actually cut) help in relating the different restriction fragments to each other. The resulting information can be used to construct a restriction map, a linear map of the relative positions of recognition sites for a variety of restriction endonucleases.

If the restriction maps of two independently isolated DNA fragments show extensive sharing of restriction sites, it is highly likely that the two fragments contain overlapping DNA sequences or are closely related members of a repeated DNA sequence family. Restriction mapping of a recombinant DNA clone immediately provides details of the length of the insert DNA and some information on the location of unique restriction sites which may be useful for subcloning purposes (where fragments of a recombinant DNA are cloned into a different vector molecule, often so as to study some aspect of its structure or function).

Vector systems for cloning different sizes of DNA fragments

Cell-based DNA cloning has been used widely as a tool for producing quantities of pure DNA for physical characterization and functional studies of individual genes, gene clusters or other DNA sequences of interest. However, the size of different DNA sequences of interest can vary enormously (e.g. human gene sizes are known to vary between 0.1 kb and 2.5 Mb).

The first cell-based cloning systems to be developed could clone only rather small DNA fragments. Recently, however, there have been rapid developments in cloning systems that permit cloning of very large DNA fragments.

Plasmid vectors provide a simple way of cloning small DNA fragments in bacterial (and simple eukaryotic) cells

In order to adapt natural plasmid molecules as cloning vectors, several modifications are normally made:

- Insertion of a multiple cloning site polylinker. This is a short (~30 bp) synthetic sequence which contains unique restriction sites for a variety of common restriction nucleases (pre-existing restriction sites for these enzymes will be deleted from the plasmid if necessary to ensure the presence of unique cloning sites).
- Insertion of an antibiotic resistance gene. The host cells that are used must naturally be sensitive to the antibiotic in question so that any vector molecule which transforms a host cell can confer antibiotic resistance. By plating transformed cells on a medium containing the antibiotic, only those cells that have been transformed by vector molecules survive.
- Insertion of a selection system for screening for recombinants. Typically this involves arranging for the multiple cloning site polylinker to be inserted into an expressible gene or gene fragment within the plasmid.

The plasmid vector pUC19 contains a polylinker with unique cloning sites for multiple restriction nucleases and an ampicillin resistance gene to permit identification of transformed cells.

In addition, selection for recombinants is achieved by insertional inactivation of a component of the β-galactosidase gene, a complementary portion of this gene being provided by using a specially modified *E. coli* host cell.

Lambda and cosmid vectors provide an efficient means of cloning moderately large DNA fragments in bacterial cells

The major disadvantage of plasmid vectors is that their capacity for accepting large DNA fragments is severely limited: most inserts are a few kilobases in length and inserts larger than 5–10 kb are very rare. Additionally, standard methods of transformation of bacterial cells with plasmid vectors are relatively inefficient. To address these difficulties, attention was focused at an early stage on the possibility of using bacteriophage lambda as a cloning vector. The wild-type λ virus particle (virion) contains a genome of close to 50 kb of linear double-stranded DNA packaged within a protein coat and has evolved a highly efficient mechanism of infecting *E. coli* cells. After the λ virion attaches to the bacterial cell, the coat protein is discarded and the λ DNA is injected into the cell. At the extreme termini of the λ DNA are overhanging 5' ends which are 12 nucleotides long and complementary in base sequence. Because these large 5' overhangs can base-pair, they are effectively sticky ends, similar to, but more cohesive than, the small sticky ends generated by some restriction nucleases. Such cohesive properties are recognized in the name given to this sequence - the cos sequence. Once inside the bacterial cell, the cos sequences base-pair, and sealing of nicks by cellular ligases results in the formation of a double-stranded circular DNA. Thereafter the λ DNA can enter one of two alternative pathways:

- The lytic cycle. The λ DNA replicates, initially bidirectionally, and subsequently by a rolling circle model which generates linear multimers of the unit length. Coat proteins are synthesized and the λ multimers are snipped at the cos sites to generate unit lengths of λ genome which are packaged within the protein coats. Some of the λ gene products lyse the host cell, allowing the virions to escape and infect new cells.
- The lysogenic state. The λ genome possesses a gene *att* which has a homolog in the *E. coli* chromosome. Apposition of the two *att* genes can result in

recombination between the λ and *E. coli* genomes and subsequent integration of the λ DNA within the *E. coli* chromosome. In this state, the λ DNA is described as a provirus and the host cell as a lysogen because, although the λ DNA can remain stably integrated for long periods, it has the capacity for excision from the host chromosome and entry into the lytic cycle. Genes required for lysogenic function are located in a central segment of the λ genome.

The decision to enter the lytic cycle or the lysogenic state is controlled by two regulatory genes, *c*I and *cro*. These two genes are mutually antagonistic: in the lytic state the *cro* protein dominates, leading to repression of *c*I, whereas in the lysogenic state the *c*I repressor dominates and suppresses transcription of other λ genes including *cro*. In normally growing host cells, the lysogenic state is favored and the λ genome replicates along with the host chromosomal DNA. Damage to host cells favors a transition to the lytic cycle, enabling the virus to escape the damaged cell and infect new cells.

In order to design suitable cloning vectors based on λ, it was necessary to design a system whereby foreign DNA could be attached to the λ replicon *in vitro* and for the resultant recombinant DNA to be able to transform *E. coli* cells at high efficiency. The latter requirement was achieved by developing an *in vitro* packaging system which mimicked the way in which wild-type λ DNA is packaged in a protein coat, resulting in high infection efficiency.

Several major types of cloning vector that have been developed by modifying phage λ, or utilizing the size selection imposed by cos sequences, are described in the following sections.

Replacement λ vectors: Only DNA molecules from 37 to 5' kb in length can be stably packaged into the λ particle. The central segment of the λ genome contains genes that are required for the lysogenic cycle but are not essential for lytic function. As a result, it can be removed and replaced by a foreign DNA fragment. Using this strategy, it is possible to clone foreign DNA up to 23 kb in length, and such vectors are normally used for making genomic DNA libraries.

Insertion λ vectors: Lambda vectors used for making cDNA libraries do not require a large insert capacity (most cDNAs are <5 kb long). Design of insertion vectors often involves modification of the λ genome to permit insertional cloning into the *c*I gene.

Cosmid vectors: Cosmid vectors contain cos sequences inserted into a small plasmid vector. Large (~30–44 kb) foreign DNA fragments can be cloned using such vectors in an *in vitro* packaging reaction because the total size of the cosmid vector is usually only about 8 kb.

Large DNA fragments can be cloned in bacterial cells using vectors based on bacteriophage P1 and F factor plasmids

Because the human and other mammalian genomes are so large, and because many individual genes can be very large, there was a need for the development of new cloning vectors that could accept large DNA inserts. A number of such vectors have been developed recently and have found immediate uses in general physical mapping of genomes and in permitting the characterization and expression of large genes or gene complexes. Examples of such vectors are discussed below.

Bacterial artificial chromosome (BAC) vectors: Many vectors which are popularly used for DNA cloning in bacterial cells contain high to medium copy number replicons. The advantage of vectors which contain such replicons is the high yield of DNA they afford: each cell in which a vector molecule is propagated will have several to multiple copies of the vector molecule, depending on the replication efficiency of its replicon. However, an important disadvantage is that such vectors often show structural instability of inserts, resulting in deletion or rearrangement of portions of the cloned DNA. Such instability is particularly common in the case of DNA inserts of eukaryotic origin where repetitive sequences occur frequently and, as a result, it is difficult to clone and maintain intact large DNA in bacterial cells.

In order to overcome this limitation, attention has recently been focused on vectors based on low copy number replicons, such as the *E. coli* fertility plasmid, the F-factor. This plasmid contains two genes, *parA* and *parB*, which maintain the copy number of the F factor at 1–2 per *E. coli* cell. Vectors based on the F factor system are able to accept large foreign DNA fragments (>300 kb). The resulting recombinants can be transferred with considerable efficiency into bacterial cells using electroporation (a method of exposing cells to high voltages in order to relax the selective permeability of their plasma membranes). However, because the resulting bacterial artificial chromosomes (BACs) contain a low copy number replicon, only very low yields of recombinant DNA can be recovered from the host cells (Shizuya *et al.*, 1992).

Bacteriophage P1 vectors and PACs: Certain bacteriophages have relatively large genomes, thereby affording the potential for developing vectors that can accommodate large foreign DNA fragments. One such is bacteriophage P1 which, like phage λ, packages its genome in a protein coat, and 110–115 kb of linear DNA is packaged in the P1 protein coat. P1 cloning vectors have therefore been designed in which components of P1 are included in a circular plasmid. The P1 plasmid vector can be cleaved to generate two vector arms to which up to 100 kb of foreign DNA can be ligated and packaged into a P1 protein coat *in vitro*. The recombinant P1 phage can be allowed to adsorb to a suitable host, following which the recombinant P1 DNA is injected into the cell, circularizes and can be amplified (Sternberg, 1992).

An improvement on the size range of inserts accepted by the basic P1 cloning system has been the use of bacteriophage T4 *in vitro* packaging systems with P1 vectors which enables the recovery of inserts up to 122 kb in size. More recently, features of the P1 and F-factor systems have been combined to produce P1-derived artificial chromosome (PAC) cloning systems (Iouannou et al., 1994).

Yeast artificial chromosomes (YACs) enable cloning of megabase fragments

The most popularly used system for cloning very large DNA fragments involves the construction of yeast artificial chromosomes (YACssee; Burke et al., 1987; Schlessinger, 1990).

Cloning in yeast cells offers, in principle, some advantages over cloning in bacterial cells. Certain eukaryotic sequences, notably those with repeated sequence organizations, are difficult, or impossible, to propagate in bacterial cells which do not have such types of DNA organization, but would be anticipated to be tolerated in yeast cells which are eukaryotic cells. However, the main advantage offered by YACs has been the ability to clone very large DNA fragments. Such a development proceeded from the realization that the great bulk of the DNA in a chromosome is not required for normal chromosome function.

i. Centromeres, required for disjunction of sister chromatids in mitosis and of homologous chromosomes at the first meiotic division.
ii. Telomeres, required for complete replication of linear molecules and for protection of the ends of the chromosome from nuclease attack.
iii. Autonomous replicating sequence (ARS) elements, required for autonomous replication of the chromosomal DNA. They are thought to act as specific replication origins.

In each case, the DNA segment necessary for functional activity *in vivo* in yeast is limited to at most a few hundred base pairs of DNA. As a result, it became possible to envisage a novel cloning system based on the use of chromosomal replicons (ARS elements) as an alternative to the ubiquitous use of cloning vectors based on extrachromosomal replicons (those found in plasmids and bacteriophages).

To make a YAC, it is simply necessary to combine four short sequences that can function in yeast cells: two telomeres, one centromere and one ARS element, together with a suitably sized foreign DNA fragment to give a linear DNA molecule in which the telomere sequences are correctly positioned at the termini. The resulting construct cannot be transfected directly into yeast cells. Instead, yeast cells have to be treated in such a way as to remove the external cell walls. The resulting yeast spheroplasts can accept exogenous fragments but are osmotically unstable and need to be embedded in agar. The overall transformation efficiency is very low and the yield of cloned

DNA is low (about one copy per cell). Nevertheless, the capacity to clone large exogenous DNA fragments (up to 2 Mb) has made YACs a vital tool in physical mapping.

Cloning systems for preparing single-stranded DNA and for studying gene expression

Single-stranded DNA clones for use in DNA sequencing are generally obtained using M13 or phagemid vectors

DNA templates provided in the form of isolated single-stranded DNA are preferred for conventional DNA sequencing because the sequences obtained are clearer and easier to read. Although single-stranded DNA templates can be prepared by PCR-based methods, single-stranded recombinant DNA clones are usually used as templates. They are obtained using vectors based on certain bacteriophages which naturally assume a single-stranded DNA form at some stage in their life cycle. Because the vector sequence is already known, it is often convenient to use a single vector-specific sequencing primer which is complementary to a sequence in the vector adjacent to the cloning site.

M13 vectors: M13 is one of a group of filamentous bacteriophages (including the fd and f1 phages) which can infect certain strains of *E. coli*. The genomes of these phages consist of single-stranded circular DNA molecules about 6.4 kb long, which are very highly related to each other in DNA sequence and enclosed in a protein coat, forming a long filamentous structure. After adsorption to the bacterium, the phage genome enters the bacterial cell where it is converted from the single-stranded form to a double-stranded form, the replicative form (RF). The latter serves as a template for making numerous copies of the genome and, after a certain time, a phage-encoded product switches DNA synthesis towards production of single strands. The latter migrate to the cell membrane where they are enclosed in a protein coat, and hundreds of mature phage particles are extruded from the infected cell without cell lysis. M13 vectors are based upon the double-stranded RF form and have a multiple cloning site for generating double-stranded recombinant DNA circles. The latter can be transfected into suitable strains of *E. coli*. After a certain period, phage particles are harvested and stripped of their protein coats to release single-stranded recombinant DNA for direct use as templates in DNA sequencing reactions.

Phagemid vectors: Phagemid vectors are plasmids which have been artificially manipulated so as to contain a small segment of the genome of a filamentous phage, such as M13, fd or f1. The selected phage sequences contain all the *cis*-acting elements required for DNA replication and assembly into phage particles. They permit successful cloning of inserts several kilobases long (unlike M13 vectors in which such inserts tend to be unstable). Following transformation of a suitable *E. coli* strain with a recombinant phagemid, the bacterial cells are *superinfected* with a filamentous helper phage, such as f1, which is required to provide the coat protein. Phage particles secreted from the superinfected cells will be a mixture of helper phage and recombinant phagemids. The mixed single-stranded DNA population can be used directly for DNA sequencing because the primer for initiating DNA strand synthesis is designed to bind specifically to a sequence of the phagemid vector adjacent to the cloning site. Commonly used phagemid vectors include the pEMBL series of plasmids and the pBluescript family.

Expression cloning involves cloning gene sequences using a cloning system which is designed to permit expression of the gene

For many purposes, DNA cloning is geared simply towards amplifying the introduced DNA to obtain sufficient quantities for a variety of subsequent structural and functional studies. However, in the case of gene sequences, there are many circumstances where instead of just amplifying and propagating the cloned DNA, it is desirable to be able to express the gene in some way (expression cloning). In each case, appropriate expression signals need to be provided by the cloning system. Expression cloning can be conducted using PCR-based systems, but is often conducted using cell-based cloning systems. A wide variety of cloning systems can be used, depending on:

- Type of expression product. For some purposes, it may be sufficient to be able to obtain an RNA product. Examples include generation of antisense RNA probes for use in tissue *in situ* hybridization studies, or generation of antisense RNA for inhibiting or destroying expression of specific genes, either in functional studies or for therapeutic purposes. In many cases, however, a protein product is desired.
- Type of environment. Sometimes, it may be sufficient to express the product *in vitro*. Often, however, it may be desirable to be able to express the product in a particular cellular system which may be a highly defined prokaryotic or eukaryotic cell line.
- Purpose of expression system. The expression system may be designed simply for investigating expression. In some cases, however, the purpose may be to retrieve large quantities of an expression product, as in the need to generate large quantities of a specific protein to assist subsequent crystallography studies or attempts to raise specific antibodies against the protein.

A large variety of expression cloning vectors have been designed to be used in different host cell systems ranging from bacterial cells to mammalian cells, with specific vectors being engineered to be useful in specific host cell types.

cDNA expression libraries for immunological screening: Expression of eukaryotic proteins in bacterial cells can also be used as a target system for screening with antibodies to identify previously uncharacterized genes. In such cases, even although there may be no information concerning the coding sequence of the gene in question, partial purification of a protein product may have allowed a specific antibody to be raised.

Using vectors such as λ gt11, it is possible to prepare a cDNA library from a desired eukaryotic tissue in which the bacterial clones can express proteins from their eukaryotic inserts. Filters containing individual bacterial colonies can be screened by exposure to the antibody. Positively reacting bacteria can then be propagated to isolate the cDNA clone, and the isolated cDNA clone can, in turn, be used to screen a genomic library to identify the cognate gene.

Large scale expression of eukaryotic polypeptides in bacterial cells: In this case, the inserted eukaryotic cDNA is simply a provider of genetic information required to specify a polypeptide; the control of expression is provided externally. Expression of important eukaryotic genes in bacterial cells can provide an endless bulk supply of important, medically relevant compounds or of proteins for basic follow-up research studies. As the production of very large amounts of a protein (which are often foreign to the host cell) can be detrimental to host cell growth, expression systems are often designed with inducible promoters. In such cases, cells with the required recombinant DNA can be selected and grown up in large quantities without expression of the foreign gene. Expression of the inserted gene can be switched on by exposure to an inducing agent and cells can be harvested shortly thereafter. One major problem in some cases, however, concerns the importance of post-translational processing in the production of many proteins. Because of differences between such systems in prokaryotes and eukaryotes, eukaryotic proteins produced by expression of cDNA clones in bacterial cells may be unstable, or show limited or no biological activity.

An important application of this approach is designed to permit the construction of an antibody specific for a eukaryotic polypeptide where none have previously been available. In these cases, the object is to prepare fusion proteins consisting of an amino terminal peptide encoded by a portion of a bacterial gene, such as the *E. coli lacZ* gene, attached to the desired eukaryotic polypeptide.

Expressing eukaryotic genes in eukaryotic cells: The biological properties of many eukaryotic proteins synthesized in bacteria may not be very representative of the native molecules because of incorrect or inefficient protein folding and post-translational processing. As a result, considerable effort has been developed to express mammalian proteins in eukaryotic and preferably mammalian cells. Major expression systems include:

- those that are designed to produce transient or stable expression of the transfected DNA. The most widely used of the transient expression systems is the monkey *COS* cell system which was derived by transformation of simian kidney CV-1 cells with a replication origin-defective SV40 genome (Gluzman, 1981);
- those that involve the use of viral expression vectors, including vectors derived from SV40, adenovirus, vaccinia virus, retroviruses and the insect baculovirus. The latter expression system is popularly used for bulk preparation of a specific eukaryotic protein.

NUCLEIC ACID HYBRIDIZATION ASSAYS

Nucleic acid hybridization is a fundamental tool in molecular genetics which takes advantage of the ability of individual single-stranded nucleic acid molecules to form double-stranded molecules (that is, to hybridize to each other). For this to happen, the interacting single-stranded molecules must have a sufficiently high degree of base complementarity. Standard nucleic acid hybridization assays involve using a labeled nucleic acid probe to identify related DNA or RNA molecules (that is, ones with a significantly high degree of sequence similarity) within a complex mixture of unlabeled nucleic acid molecules, the target nucleic acid.

Preparation of nucleic acid probes

In standard nucleic acid hybridization assays the probe is labeled in some way. Nucleic acid probes may be made as single-stranded or double-stranded molecules (see *Figure 5.1*), but the working probe must be in the form of single strands.

Conventional DNA probes are isolated by cell-based DNA cloning or by PCR. In the former case, the starting DNA may range in size from 0.1 kb to hundreds of kilobases in length and is usually (but not always) originally double-stranded. PCR-derived DNA probes have often been less than 10 kb long and are usually, but not always, originally double-stranded. Conventional DNA probes are usually labeled by incorporating labeled dNTPs during an *in vitro* DNA synthesis reaction.

RNA probes can conveniently be generated from DNA which has been cloned in a specialized plasmid vector (Melton *et al.*, 1984). Such vectors normally contain a phage promoter sequence immediately adjacent to the multiple cloning site. An RNA synthesis reaction is employed using the relevant phage RNA polymerase and the four rNTPs, at least one of which is labeled. Specific labeled RNA transcripts can then be generated from the cloned insert.

Oligonucleotide probes are short (typically 15–50 nucleotides) single-stranded pieces of DNA made by chemical synthesis: mononucleotides are added, one at a time, to a starting mononucleotide, conventionally the 3' end nucleotide, which is bound to a solid support. Generally, oligonucleotide probes are designed with a specific sequence chosen in response to prior information about the target DNA. Sometimes, however, oligonucleotide probes are used which are degenerate in sequence. Typically this involves parallel syntheses of a set of oligonucleotides which are identical at certain nucleotide positions but different at others. Oligonucleotide probes are often labeled by incorporating a ^{32}P atom or other labeled group at the 5' end.

DNA and RNA can conveniently be labeled in vitro by incorporation of nucleotides (or nucleotide components) containing a labeled atom or chemical group

Although, in principle, DNA and RNA can be labeled *in vivo*, by supplying labeled deoxynucleotides to tissue culture cells, this procedure is of limited general use; it has been restricted largely to preparing labeled viral DNA from virus-infected cells, and studying RNA processing events. A much more versatile method involves *in vitro* labeling: the purified DNA, RNA or oligonucleotide is labeled *in vitro* by using a suitable enzyme to incorporate labeled nucleotides. Two major types of procedure have been widely used:

- Labeling of new strands during*in vitro*DNA or RNA synthesis. In this type of procedure, DNA or RNA polymerase is used to make labeled DNA or RNA copies of a starting DNA. The *in vitro* DNA or RNA synthesis reaction requires that at least one of the four nucleotide precursors carries a labeled group. Labeling of DNA by *in vitro* DNA synthesis is normally accomplished using one of three methods: nick-translation (this section); random primed labeling (this section); or PCR-mediated labeling. Labeling of RNA generally is carried out using an *in vitro* transcription system.

- End-labeling. This type of procedure involves addition of a labeled group to one or a few terminal nucleotides. It is less widely used, but is useful for a number of procedures, including labeling of single-stranded oligonucleotides (see below) and restriction mapping. Inevitably, because only one or a very few labeled groups are incorporated, the specific activity (the amount of radioactivity incorporated divided by the total mass) of the labeled DNA is much less than that for probes in which there has been incorporation of several labeled nucleotides along the length of the DNA.

Labeling DNA by nick translation: The nick-translation procedure involves introducing single-strand breaks (*nicks*) in the DNA, leaving exposed 3' hydroxyl termini and 5' phosphate termini. The nicking can be achieved by adding a suitable endonuclease such as pancreatic deoxyribonuclease I (DNase I). The exposed nick can then serve as a start point for introducing new nucleotides at the 3' hydroxyl side of the nick using the DNA polymerase activity of *E. coli* DNA polymerase I at the same time as existing nucleotides are removed from the other side of the nick by the 5' $\rightarrow$ 3' exonuclease activity of the same enzyme. As a result, the nick will be moved progressively along the DNA ('translated') in the 5' $\rightarrow$ 3' direction. If the reaction is carried out at a relatively low temperature (about 15° C), the reaction proceeds no further than one complete renewal of the existing nucleotide sequence. Although there is no net DNA synthesis at these temperatures, the synthesis reaction allows the incorporation of labeled nucleotides in place of the previously existing unlabeled ones.

Random primed DNA labeling: The random primed DNA labeling method (sometimes known as oligolabeling) (Feinberg and Vogelstein, 1983) is based on hybridization of a mixture of all possible hexanucleotides: the starting DNA is denatured and then cooled slowly so that the individual hexanucleotides can bind to suitably complementary sequences within the DNA strands. Synthesis of new complementary DNA strands is primed by the bound hexanucleotides and is catalyzed by the Klenow subunit of DNA polymerase I (which contains the polymerase activity in the absence of associated exonuclease activities). DNA synthesis occurs in the presence of the four dNTPs, at least one of which has a labeled group. This method produces labeled DNAs of high specific activity. Because all sequence combinations are represented in the hexanucleotide mixture, binding of primer to template DNA occurs in a random manner, and labeling is uniform across the length of the DNA.

End-labeling of DNA: Single-stranded oligonucleotides are usually end-labeled using polynucleotide kinase (kinase end-labeling). Typically, the label is provided in the form of a ^{32}P at the γ-phosphate position of ATP and the polynucleotide kinase catalyses an exchange reaction with the 5' -terminal phosphates. The same procedure can also be used for labeling double-stranded DNA. In this case, fragments carrying label at one end only can then be generated by cleavage at an internal restriction site, generating two differently sized fragments which can be separated by gel electrophoresis and purified.

Larger DNA fragments can be end-labeled by various alternative methods. Fill-in end-labeling is one popular approach, and uses the *Klenow subunit* of *E. coli* DNA polymerase. Again, fragments carrying label at one end only can be generated by restriction cleavage and size fractionation. An alternative PCR-based method is primer-mediated 5' end-labeling.

Labeling of RNA: The preparation of labeled RNA probes (riboprobes) is most easily achieved by *in vitro* transcription of insert DNA cloned in a suitable plasmid vector. The vector is designed so that adjacent to the multiple cloning site is a phage promoter sequence, which can be recognized by the corresponding phage RNA polymerase. For example, the plasmid vector pSP64 contains the bacteriophage SP6 promoter sequence immediately adjacent to a multiple cloning site. The SP6 RNA polymerase can then be used to initiate transcription from a specific start point in the SP6 promoter sequence, transcribing through any DNA sequence that has been inserted into the multiple cloning site. By using a mix of NTPs, at least one of which is labeled, high specific activity radiolabeled transcripts can be generated. Bacteriophage T3 and T7 promoter/RNA polymerase systems are also used commonly for generating riboprobes. Labeled sense and antisense riboprobes can be generated from any gene cloned in such vectors (the gene can be cloned in either of the two orientations) and are widely used in tissue *in situ* hybridization.

Nucleic acids can be labeled by isotopic and nonisotopic methods

Isotopic labeling and detection: Traditionally, labeling of nucleic acids has been conducted by incorporating nucleotides containing radioisotopes. Such radiolabeled probes contain nucleotides with a radioisotope (often ^{32}P, ^{33}P, ^{35}S or ^{3}H), which can be detected specifically in solution or, much more commonly, within a solid specimen.

The intensity of an autoradiographic signal is dependent on the intensity of the radiation emitted by the radioisotope, and the time of exposure, which may often be long (one or more days, or even weeks in some applications). ^{32}P has been used widely in Southern blot hybridization, dot-blot hybridization, colony and plaque hybridization because it emits high energy β-particles which afford a high degree of sensitivity of detection. It has the disadvantage, however, that it is relatively unstable. Additionally, its high energy β-particle emission can be a disadvantage under circumstances when fine physical resolution is required to interpret the resulting image unambiguously. For this reason, radionuclides which provide less energetic β-particle radiation have been preferred in certain procedures, for example ^{35}S-labeled and ^{33}P-labeled nucleotides for DNA sequencing and tissue *in situ* hybridization, and ^{3}H-labeled nucleotides for chromosome *in situ* hybridization. ^{35}S and ^{33}P have moderate half-lives while ^{3}H has a very long half-life. However, the latter isotope is disadvantaged by its comparatively low energy β-particle emission which necessitates very long exposure times.

^{32}P-labeled and ^{33}P-labeled nucleotides used in DNA strand synthesis labeling reactions have the radioisotope at the α-phosphate position, because the β- and γ-phosphates from dNTP precursors are not incorporated into the growing DNA chain. Kinase-mediated end-labeling, however, uses [γ-^{32}P]ATP. In the case of ^{35}S-labeled nucleotides which are incorporated during the synthesis of DNA or RNA strands, the NTP or dNTP carries a ^{35}S isotope in place of the O^- of the α-phosphate group. ^3H-labeled nucleotides carry the radioisotope at several positions. Specific detection of molecules carrying a radioisotope is most often performed by autoradiography.

Nonisotopic labeling and detection: Nonisotopic labeling systems involve the use of nonradioactive probes. Although developed only comparatively recently, they are becoming increasingly popular and are finding increasing applications in a variety of different areas (Kricka, 1992). Two types of non-radioactive labeling are conducted:

- Direct nonisotopic labeling, where a nucleotide which contains the label that will be detected is incorporated. Often such systems involve incorporation of modified nucleotides containing a fluorophore, a chemical group which can fluoresce when exposed to light of a certain wavelength.
- Indirect nonisotopic labeling, usually featuring the chemical coupling of a modified reporter molecule to a nucleotide precursor. After incorporation into DNA, the reporter groups can be specifically bound by an affinity molecule, a protein or other ligand which has a very high affinity for the reporter group. Conjugated to the latter is a marker molecule or group which can be detected in a suitable assay. The reporter molecules on modified nucleotides need to protrude sufficiently far from the nucleic acid backbone to facilitate their detection by the affinity molecule and so a long carbon atom *spacer* is required to separate the nucleotide from the reporter group.

Two indirect nonisotopic labeling systems are widely used:

- The biotin-streptavidin system utilizes the extremely high affinity of two ligands: biotin (a naturally occurring vitamin) which acts as the reporter, and the bacterial protein streptavidin, which is the affinity molecule. Biotin and streptavidin bind together extemely tightly with an affinity constant of 10^{-14}, one of the strongest known in biology. Biotinylated probes can be made easily by including a suitable biotinylated nucleotide in the labeling reaction.
- Digoxigenin is a plant steroid (obtained from *Digi-talis* plants) to which a specific antibody has been raised. The digoxigenin-specific antibody permits detection of nucleic acid molecules which have incorporated nucleotides containing the digoxigenin reporter group.

A variety of different marker groups or molecules can be conjugated to affinity molecules such as streptavidin or the digoxigenin-specific antibody. They include various fluorophores, or enzymes such as alkaline phosphatase and peroxidase which can permit detection via colorimetric assays or chemical luminescence assays, etc.

Principles of nucleic acid hybridization

Nucleic acid hybridization is a method for identifying closely related nucleic acid molecules within two populations, a complex target population and a comparatively homogeneous probe population

Definition of nucleic acid hybridization: Nucleic acid hybridization involves mixing single strands of two sources of nucleic acids, a probe which typically consists of a homogeneous population of *identified molecules* (e.g. cloned DNA or chemically synthesized oligonucleotides) and a target which typically consists of a *complex, heterogeneous population* of nucleic acid molecules. If either the probe or the target is initially double-stranded, the individual strands must be separated, generally by heating or by alkaline treatment. After mixing single strands of probe with single strands of target, strands with complementary base sequences can be allowed to reassociate.

Complementary probe strands can reanneal to form *homoduplexes*, as can complementary target DNA strands. However, it is the annealing of a probe DNA strand and a complementary target DNA strand to form a labeled probe-target heteroduplex that defines the usefulness of a nucleic acid hybridization assay. The rationale of the hybridization assay is to use the identified probe to query the target DNA by identifying fragments in the complex target which may be related in sequence to the probe.

Melting temperature and hybridization stringency: Denaturation of double-stranded probe DNA is generally achieved by heating a solution of the labeled DNA to a temperature which disrupts the hydrogen bonds that hold the two complementary DNA strands together. The energy required to separate two perfectly complementary DNA strands is dependent on a number of factors, notably:

- strand length - long homoduplexes contain a large number of hydrogen bonds and require more energy to separate them; because the labeling procedure typically results in short DNA probes, this effect is negligible above an original length (i.e. prior to labeling) of 500 bp;
- base composition - because GC base pairs have one more hydrogen bond than AT base pairs, strands with a high % GC composition are more difficult to separate than those with a low % GC composition;
- chemical environment - the presence of monovalent cations (e.g. Na^+ ions) stabilizes the duplex, whereas chemical denaturants such as formamide and urea destabilize the duplex by chemically disrupting the hydrogen bonds.

A useful measure of the stability of a nucleic acid duplex is the melting temperature (T_m). This is the temperature corresponding to the midpoint in the observed transition from double-stranded to single-stranded form. Conveniently, this transition can be followed by measuring the optical density of the DNA. The bases of the nucleic acids absorb 260 nm ultraviolet (UV) light strongly. However, the adsorption by double-stranded

DNA is considerably less than that of the free nucleotides. This difference, the so-called hypochromic effect, is due to interactions between the electron systems of adjacent bases, arising from the way in which adjacent bases are stacked in parallel in a double helix. If duplex DNA is gradually heated, therefore, there will be an increase in the light absorbed at 260 nm (the optical density$_{260}$ or OD_{260}) towards the value characteristic of the free bases. The temperature at which there is a midpoint in the optical density shift is then taken as the T_m.

For mammalian genomes, with a base composition of about 40% GC, the DNA denatures with a T_m of about 87°C under approximately physiological conditions. The T_m of perfect hybrids formed by DNA, RNA or oligonucleotide probes can be determined according to the formulae7. Often, hybridization conditions are chosen so as to promote heteroduplex formation and the hybridization temperature is often as much as 25°C below the T_m.

However, after the hybridization and removal of excess probe, hybridization washes may be conducted under more stringent conditions so as to disrupt all duplexes other than those between very closely related sequences. Probe-target heteroduplexes are most stable thermodynamically when the region of duplex formation contains perfect base matching. Mismatches between the two strands of a heteroduplex reduce the T_m: for normal DNA probes, each 1% of mismatching reduces the T_m by approximately 1°C. Although probe-target heteroduplexes are usually not as stable as reannealed probe homoduplexes, a considerable degree of mismatching can be tolerated if the overall region of base complementarity is long.

Increasing the concentration of NaCl and reducing the temperature reduces the hybridization stringency, and enhances the stability of mismatched heteroduplexes. This means that comparatively diverged members of a multigene family or other repetitive DNA family can be identified by hybridization using a specific family member as a probe. Additionally, a gene sequence from one species can be used as a probe to identify homologs in other comparatively diverged species, provided the sequence is reasonably conserved during evolution.

Conditions can also be chosen to maximize hybridization stringency (e.g. lowering the concentration of NaCl and increasing the temperature), so as to encourage dissociation (denaturation) of mismatched heteroduplexes. If the region of base complementarity is small, as with oligonucleotide probes (typically 15–20 nucleotides), hybridization conditions can be chosen such that a single mismatch renders a heteroduplex unstable.

The kinetics of DNA reassociation are defined by the product of DNA concentration and time (C_0t)

When double-stranded DNA is denatured, for example by heat, and the complementary single strands are then allowed to reassociate to form double-stranded DNA, the speed at which the complementary strands reassociate will depend on the starting concentration of the DNA. If there is a high concentration of the complementary DNA sequences, the time taken for any one single-stranded DNA molecule to find a complementary strand and form a duplex will be reduced. Reassociation kinetics is the term used to measure the speed at which complementary single-stranded molecules are able to find each other and form duplexes. It is determined by two major parameters: the starting concentration (C_0) of the specific DNA sequence in moles of nucleotides per liter and the reaction time (t) in seconds. Since the rate of reassociation is proportional to C_0 and to t, the C_0t value (often loosely referred to as the cot value) is a useful measure. The C_0t value will also vary depending on the temperature of reassociation and the concentration of monovalent cations. As a result, it is usual to use fixed reference values: a reassociation temperature of 65°C and a Na^+ concentration of 0.3 M NaCl.

Most hybridization assays use an excess of target nucleic acid over probe in order to encourage probe- target formation. This is so because the probe is usually *homogenous*, often consisting of a single type of cloned DNA molecule or RNA molecule, but the target nucleic acid is typically *heterogeneous*, comprising for example genomic DNA or total cellular RNA. In the latter case the concentration of any one sequence may be very low, thereby causing the rate of reassociation to be slow.

For example, if a Southern blot uses a cloned β-globin gene as a probe to identify complementary sequences in human genomic DNA, the latter will be present in very low concentration (the β-globin gene is an example of a single copy sequence and in this case represents only 0.00005% of human genomic DNA). It is therefore necessary to use several micrograms of target DNA to drive the reaction. By contrast, certain other sequences are highly repeated in genomic DNA, and this greatly elevated DNA concentration results in a comparatively rapid reassociation time.

Because the amount of target nucleic acid bound by a probe depends on the copy number of the recognized sequence, hybridization signal intensity is proportional to the copy number of the recognized sequence. Single copy genes give weak hybridization signals, highly repetitive DNA sequences give very strong signals. If a particular probe is heterogeneous and contains a low copy sequence of interest, such as a specific gene, mixed with a highly abundant DNA repeat, the weak hybridization signal obtained with the former will be completely masked by the strong repetitive DNA hybridization signal. This effect can, however, be overcome by competition hybridization.

A wide variety of nucleic acid hybridization assays can be used

Early experiments in nucleic acid hybridization utilized solution hybridization, involving mixing of aqueous solutions of probe and target nucleic acids. However, the very low concentration of single copy sequences in complex genomes meant that reassociation times were inevitably slow. One widely used way of increasing the reassociation speed is to artificially increase the overall DNA concentration in aqueous solution by abstracting water molecules (e.g. by adding high concentrations of polyethylene glycol).

An alternative to solution hybridization which facilitated detection of reassociated molecules involved immobilizing the target DNA on a solid support, such as a membrane made of nitrocellulose or nylon, to both of which single-stranded DNA binds readily. Attachment of labeled probe to the immobilized target DNA can then be followed by removing the solution containing unbound probe DNA, extensive washing and drying in preparation for detection.

This is the basis of the standard nucleic acid hybridization assays currently in use. More recently, however, reverse hybridization assays have become more popular. In these cases, the probe population is unlabeled and fixed to the solid support, while the target nucleic acid is labeled and present in aqueous solution. Note, therefore, that probe and target are not primarily distinguished by which is the labeled and which is the unlabeled population. Instead, the important consideration is that the target DNA should be the complex imperfectly understood population which the probe (whose molecular identity is known) attempts to query. Depending on the nature and form of the probe and target, a very wide variety of nucleic acid hybridization assays can be devised.

Nucleic acid hybridization assays using cloned DNA probes to screen uncloned nucleic acid populations

Numerous applications in molecular genetics involve taking an individual DNA clone and using it as a hybridization probe to screen for the presence of related sequences within a complex target of uncloned DNA or RNA. Sometimes the assay is restricted to simply checking for presence or absence of sequences related to the probe. In other cases, useful information can be obtained regarding the size of the complementary sequences, their subchromosomal location or their locations within specific tissues or groups of cells.

Dot-blot hybridization is a rapid screening method which often employs allele-specific oligonucleotide (ASO) probes to discriminate between alleles differing at a single nucleotide position

The general procedure of dot-blotting involves taking an aqueous solution of target DNA, for example total human genomic DNA, and simply spotting it on to a nitrocellulose or nylon membrane then allowing it to dry. The variant technique of slot-blotting involves pipetting the DNA through an individual slot in a suitable template. In both methods the target DNA sequences are denatured, either by previously exposing to heat, or by exposure of the filter containing them to alkali. The denatured target DNA sequences now immobilized on the membrane are exposed to a solution containing single-stranded labeled probe sequences.

After allowing sufficient time for probe-target heteroduplex formation, the probe solution is decanted, and the membrane is washed to remove excess probe that may have become nonspecifically bound to the filter. It is then dried and exposed to an autoradiographic film.

A useful application of dot-blotting involves distinguishing between alleles that differ by even a single nucleotide substitution. To do this allele-specific oligonucleotide (ASO) probes are constructed from sequences spanning the variant nucleotide site. ASO probes are typically 15–20 nucleotides long and are normally employed under hybridization conditions at which the DNA duplex between probe and target is stable *only if there is perfect base complementarity between them*: a single mismatch between probe and target sequence is sufficient to render the short heteroduplex unstable.

Typically, this involves designing the oligonucleotides so that the single nucleotide difference between alleles occurs in a central segment of the oligonucleotide sequence, thereby maximizing the thermodynamic instability of a mismatched duplex. Such discrimination can be employed for a variety of research and diagnostic purposes. Although ASOs can be used in conventional Southern blot hybridization, it is more convenient to use them in dot-blot assays.

Another method of ASO dot blotting uses a reverse dot-blotting approach. This means that the oligonucleotide probes are not labeled and are fixed on a filter or membrane whereas the target DNA is labeled and provided in solution. Positive binding of labeled target DNA to a specific oligonucleotide on the membrane is taken to mean that the target has that specific sequence. This approach, and related DNA microarray methods, have many diagnostic applications.

Southern and Northern blot hybridizations detect target DNA and RNA fragments that have been size-fractionated by gel electrophoresis

Southern blot hybridization: In this procedure, the target DNA is digested with one or more restriction endonucleases, size-fractionated by agarose gel electrophoresis, denatured and transferred to a nitrocellulose or nylon membrane for hybridization. During the electrophoresis, DNA fragments, which are negatively charged because of the phosphate groups, are repelled from the negative electrode towards the positive electrode, and sieved through the porous gel. Smaller DNA fragments move faster. For fragments between 0.1 and 20 kb long, the migration speed depends on fragment length, but scarcely at all on the base composition. Thus, fragments in this size range are fractionated by size in a conventional agarose gel electrophoresis system. To achieve efficient size-fractionation of large fragments (40 kb to several megabases), a more specialized system is required, such as a pulsed-field gel electrophoresis apparatus.

Following electrophoresis, the test DNA fragments are denatured in strong alkali. As agarose gels are fragile, and the DNA in them can diffuse within the gel, it is usual to transfer the denatured DNA fragments by blotting on to a durable nitrocellulose or nylon membrane, to which single-stranded DNA binds readily. The individual DNA fragments become immobilized on the membrane at positions which are a faithful record of the size separation achieved by agarose gel electrophoresis. Subsequently, the immobilized single-stranded target DNA sequences are allowed to associate with labeled single-stranded probe DNA. The probe will bind only to related DNA sequences in the target DNA, and their position on the membrane can be related back to the original gel in order to estimate their size.

An important application of Southern blot hybridization in mammalian genetics is the ability to identify a given DNA probe as a member of a repetitive DNA family. Many important mammalian genes belong to multigene families, and many other DNA sequences show varying degrees of repetition. Once a newly isolated probe is demonstrated to be related to other uncharacterized sequences, attempts can then be made to isolate the other members of the family by screening genomic DNA libraries. Additionally, screening can also be conducted on genomic DNA samples from different species to identify interspecific related sequences. An important route to identifying coding DNA involves identifying sequences that are highly conserved in evolution.

Northern blot hybridization: Northern blot hybridization is a variant of Southern blotting in which the target nucleic acid is RNA instead of DNA. A principal use of this method is to obtain information on the expression patterns of specific genes. Once a gene has been cloned, it can be used as a probe and hybridized against a Northern blot containing, in different lanes, samples of RNA isolated from a variety of different tissues. The data obtained can provide information on the range of cell types in which the gene is expressed, and the relative abundance of transcripts. Additionally, by revealing transcripts of different sizes, it may provide evidence for the use of alternative promoters, splice sites or polyadenylation sites.

Southern blot hybridization permits restriction mapping and assay of RFLPs and moderately small scale mutations

Southern blot hybridization has been used extensively in molecular genetic studies as a means of genomic restriction mapping: a labeled DNA probe from one genome can be used to infer the structure of related sequences in the same or different genomes. Because the genomic DNA samples are fractionated by separation of restriction fragments according to size, mutations which alter a restriction site, and significantly large insertions or deletions occurring between neighboring restriction sites, can be typed. Such mutations will result in altered restriction fragment lengths, that is restriction fragment length polymorphisms (RFLPs).

Direct detection of pathogenic point mutations by restriction mapping: Very occasionally, a pathogenic mutation directly abolishes or creates a restriction site, enabling direct screening for the pathogenic mutation. For example, the sickle cell mutation is a single nucleotide substitution (A → T) at codon 6 in the β-globin gene, which causes a missense mutation (Glu → Val), and at the same time abolishes an *Mst*II restriction site which spans codons 5 to 7. The nearest flanking restriction sites for *Mst*II, located 1.2 kb upstream in the 5' -flanking region and 0.2 kb downstream at the 3' end of the first intron, are well conserved. Consequently, a β-globin DNA probe can differentiate the normal β^A-globin and the mutant β^S-globin alleles in *Mst*II-digested human DNA: the former exhibits 1.2 kb and 0.2 kb *Mst*II fragments, whereas the sickle cell allele exhibits a 1.4 kb *Mst*II fragment.

Detection of conventional RFLPs: The great majority of mutations are not associated with disease; instead, they often occur within noncoding DNA sequences. As a large number of recognition sequences are known for type II restriction endonucleases, many point mutation polymorphisms will be characterized by alleles which possess or lack a recognition site for a specific restriction endonuclease and therefore display restriction site polymorphism (RSP). Accordingly, individual RSPs normally have two detectable alleles (one lacking and one possessing the specific restriction site). RSPs can be assayed by digesting genomic DNA samples with the relevant restriction endonuclease and identifying specific restriction fragments whose lengths are characteristic of the two alleles, so-called RFLPs.

VNTR-based RFLPs and DNA fingerprinting: DNA probes can also be used to monitor VNTR polymorphisms where alleles differ by a variable number of tandem repeats. To do this, genomic DNA samples are digested with a restriction endonuclease which recognizes well-conserved restriction sites flanking a specific VNTR locus. The resulting restriction fragments are separated according to size on agarose gels, transferred to a suitable membrane and hybridized with a probe representing a unique sequence from the corresponding locus. The resulting pattern of locus-specific RFLPs does not reflect RSP: instead, the differences in sizes of the restriction fragments represent integral numbers of the tandemly repeated unit.

Although the term VNTR could, in theory, encompass a wide range of repeat lengths, in practice the term is usually reserved for moderately large arrays of a repeat unit which is typically in the 5–64 bp region (so-called hypervariable minisatellite DNA, distinguishing it from simple tandem repeat polymorphism (where the repeat unit length is from 1 to 4 bp; i.e. microsatellite DNA) and tandem repeat polymorphism associated with very large arrays of satellite DNA.

If the VNTR locus is a member of a repeated DNA family, the use of a VNTR repeat probe, rather than a unique flanking probe, will produce a complex polymorphic pattern. For example, hypervariable minisatellite DNA clones have been used as probes against Southern blots of appropriately digested genomic DNA. Cross-hybridization of such probes with the other members of this highly repeated DNA family results in a pattern of hybridization bands representing the summed contributions of two alleles at each of many hypervariable loci scattered throughout the genome. Consequently, the overall polymorphism of the multilocus hybridization patterns is uniquely high. Because it permits distinction between any two individuals who are not identical twins, probing with hypervariable minisatellites has been termed DNA fingerprinting.

Detection of gene deletions by restriction mapping: Certain diseases are associated with a high frequency of deletion of all or part of a gene. If a partial restriction map has been established for the gene under investigation, deletions can be screened by Southern blot hybridization using an appropriate intragenic DNA probe. If the deletion is a small one, for example a few hundred base pairs, it is often apparent as a consistent reduction in size of normal restriction fragments in the gene. An individual who is homozygous for this mutation, or is a heterozygote with one normal allele and another with a small deletion, can easily be identified by detecting the aberrant size restriction fragments.

Large deletions will lead to absence of specific restriction fragments. Homozygous deletion of large DNA segments can easily be detected as complete absence of appropriate restriction fragments associated with the gene. If, however, an individual is heterozygous for a relatively large gene deletion, the deletion may still be detected by demonstrating comparatively reduced intensity of specific gene fragments. For example, patients with 21-hydroxylase deficiency often have deletions of about 30 kb of the 21-hydroxylase/*C4* gene cluster. Such pathological deletions eliminate the functional 21-hydroxylase gene, *CYP21*, and an adjacent *C4B* gene, leaving the related *CYP21P* pseudogene and *C4A* genes. Patients with homozygous deletions will show absence of diagnostic restriction fragments associated with *CYP21* and *C4B*, while carriers of the deletion will show a 2:1 ratio of *CYP21P*:*CYP21* and of *C4A*:*C4B*.

In situ hybridization usually involves hybridizing a nucleic acid probe to the denatured DNA of a chromosome preparation or the RNA of a tissue section fixed on a glass slide

Chromosome in situ hybridization: A simple procedure for mapping genes and other DNA sequences is to hybridize a suitable labeled DNA probe against chromosomal DNA that has been denatured *in situ*. To do this, an air-dried microscope slide preparation of metaphase or prometaphase chromosomes is made, usually from peripheral blood lymphocytes or lymphoblastoid cell lines. Treatment with RNase and proteinase K results in partially purified chromosomal DNA, which is denatured by exposure to formamide. The denatured DNA is then available for *in situ* hybridization with an added solution containing a labeled nucleic acid probe, overlaid with a coverslip. Depending on the particular technique that is used, chromosome banding of the chromosomes can be arranged either before or after the hybridization step. As a result, the signal obtained after removal of excess probe can be correlated with the chromosome band pattern in order to identify a map location for the DNA sequences recognized by the probe. Chromosome *in situ* hybridization has been revolutionized by the use of fluorescence *in situ* hybridization (FISH) techniques.

Tissue in situ hybridization: In this procedure, a labeled probe is hybridized against RNA in tissue sections (Wilkinson, 1998). Tissue sections are made from either paraffin-embedded or frozen tissue using a cryostat, and then mounted on to glass slides. A hybridization mix including the probe is applied to the section on the slide and covered with a glass coverslip. Typically, the hybridization mix has formamide at a concentration of 50% in order to reduce the hybridization temperature and minimize evaporation problems.

Although double-stranded cDNAs have been used as probes, single-stranded complementary RNA probes (riboprobes) are preferred: the sensitivity of initially single-stranded probes is generally higher than that of double-stranded probes, presumably because a proportion of the denatured double-stranded probe renatures to form probe homoduplexes. cRNA riboprobes that are complementary to the mRNA of a gene are known as antisense riboprobes and can be obtained by cloning a gene in the reverse orientation in a suitable vector such as pSP64. In such cases, the phage polymerase will synthesize labeled transcripts from the opposite DNA strand to that which is normally transcribed *in vivo*. Useful controls for such reactions include sense riboprobes which should not hybridize to mRNA except in rare occurrences where both DNA strands of a gene are transcribed.

Labeling of probes is performed using either selected radioisotopes, notably ^{35}S, or by nonisotopic labeling. In the former case, the hybridized probe is visualized using autoradiographic procedures. The localization of the silver grains is often visualized using only dark-field microscopy (direct light is not allowed to reach the objective; instead, the illuminating rays of light are directed from the side so that only scattered light enters the microscopic lenses and the signal appears as an illuminated object against a black background). However, bright-field microscopy (where the image is obtained by direct transmission of light through the sample) provides better signal detection. Fluorescence labeling is a popular nonisotopic labeling approach and detection is accomplished by fluorescence microscopy.

Nucleic acid hybridization assays using cloned target DNA, and microarray hybridization technology

Some of the technologies described in the preceding section (e.g. Southern blot hybridization and dot-blot hybridization) are also used to study cloned DNA as well as uncloned DNA. The techniques described in the next two sections, however, are dedicated to analysing cloned DNA. In addition, the very recently developed and very powerful microarray technologies are described in a third section.

Colony blot and plaque lift hybridization are methods for screening separated bacterial colonies or plaques following bacteriophage infection of bacteria

A specific approach can be used when the desired recombinant DNA contains a DNA sequence that is closely related to an available nucleic acid probe.

In the case of bacterial cells used to propagate plasmid recombinants, the cell colonies are allowed to grow on an agar surface and then transferred by surface contact to a nitrocellulose or nylon membrane, a process known as colony blotting.

Alternatively the cell mixture is spread out on a nitrocellulose or nylon membrane placed on top of a nutrient agar surface, and colonies are allowed to form directly on top of the membrane. In either approach, the membrane is then exposed to alkali to denature the DNA prior to hybridizing with a labeled nucleic acid probe.

After hybridization, the probe solution is removed, and the filter is washed extensively, dried and submitted to autoradiography using X-ray film. The position of strong radioactive signals is related back to a master plate containing the original pattern of colonies, in order to identify colonies containing DNA related to the probe. These can then be individually picked and amplified in culture prior to DNA extraction and purification of the recombinant DNA.

A similar process is possible when using phage vectors. The plaques which are formed following lysis of bacterial cells by phage will contain residual phage particles. A nitrocellulose or nylon membrane is placed on top of the agar plate in the same way as above, and when removed from the plate will constitute a faithful copy of the phage material in the plaques, a so called plaque-lift.

Gridded high density arrays of transformed cell clones or DNA clones have greatly increased the efficiency of DNA library screening

Once it became possible to create complex DNA libraries, more efficient methods of clone screening were required. Rather than simply plate out cell colonies on a cell culture dish and transfer them to a membrane in standard colony blotting, it was preferable to pick individual colonies and transfer them onto large membranes in the format of a high density gridded array.

The process was enormously simplified by the application of robotic gridding devices which could perform the necessary spotting automatically by pipetting from clones arranged in microtiter dishes into pre-determined linear coordinates on a membrane. The resulting high density clone filters permitted rapid and efficient identification library screening and could be copied and distributed to numerous laboratories throughout the world.

DNA microarray technology has enormously extended the power of nucleic acid hybridization

Recently developed DNA microarrays ('DNA chips') have provided a scale-up in hybridization assay technology because of their huge capacity for miniaturization and automation (Schena *et al.*, 1998). Although reminiscent of the filter-based arrays, microarray construction involves quite different procedures. The surfaces involved are glass rather than porous membranes and the microarrays can be divided into two main classes according to their method of construction:

- Microarrays of DNA clones delivered by microspotting. Here the DNA clones have been prepared in advance and then printed onto the surface of a microscope slide.
- Microarrays of oligonucleotides synthesized *in situ*. This approach has been pioneered by the company Affymetrix, Inc., in Santa Clara, CA, and typically involves a combination of photolithography technology from the semiconductor industry with the chemistry of oligonucleotide synthesis.

As in the case of reverse dot-blotting, the DNA microarray technologies employ a reverse nucleic acid hybridization approach: the probes consist of unlabeled DNA fixed to a solid support (the arrays of DNA or oligonucleotides) and the target is labeled and in solution.

Although the technology for establishing DNA microarrays was only developed in the last few years, already there have been numerous important applications and their impact on future biomedical research and diagnostic approaches is expected to be profound.

4

ORGANIZATION OF THE HUMAN GENOME

General organization of the human genome

The human genome is the term used to describe the total genetic information (DNA content) in human cells. It really comprises two genomes: a complex nuclear genome which accounts for 99.9995% of the total genetic information, and a simple mitochondrial genome which accounts for the remaining 0.0005%.

The nuclear genome provides the great bulk of essential genetic information, most of which specifies polypeptide synthesis on cytoplasmic ribosomes.

Mitochondria possess their own ribosomes and the few polypeptide-encoding genes in the mitochondrial genome produce mRNAs which are translated on the mitochondrial ribosomes. However, the mitochondrial genome specifies only a very small portion of the specific mitochondrial functions; the bulk of the mitochondrial polypeptides are encoded by nuclear genes and are synthesized on cytoplasmic ribosomes, before being imported into the mitochondria.

Like other complex genomes, a sizeable component of the human genome is made up of noncoding DNA. In addition, the human genome is representative of mammalian genomes and other complex genomes in having a considerable amount of repetitive DNA, including both noncoding repetitive DNA and multiple copy genes and gene fragments.

The mitochondrial genome consists of a small circular DNA duplex which is densely packed with genetic information

General structure and inheritance of the mitochondrial genome: The human mitochondrial genome is defined by a single type of circular double-stranded DNA whose complete nucleotide sequence has been established (Anderson *et al.*, 1981). It is 16 569 bp in length and is 44% (G + C). The two DNA strands have significantly different base compositions: the heavy (H) strand is rich in guanines, the light (L) strand is rich in cytosines. Although the mitochondrial DNA is principally double-stranded, a small section is defined by a triple DNA strand structure. This is because a short segment of the heavy strand is replicated for a second time, giving a structure known as 7S DNA. Human cells usually contain thousands of copies of the double-stranded mitochondrial DNA molecule. Accordingly, although a single mitochondrial DNA duplex has only about 1/8000 as much DNA as an average sized chromosome, the total mitochondrial DNA complement can account for up to about 0.5% of the DNA in a nucleated somatic cell.

During zygote formation, a sperm cell contributes its nuclear genome, but not its mitochondrial genome, to the egg cell. Consequently, the mitochondrial genome of the zygote is determined exclusively by that originally found in the unfertilized egg. The mitochondrial genome is therefore maternally inherited: males and females both inherit their mitochondria from their mother but males cannot transmit their mitochondria to subsequent generations.

Mitochondrial genes: The human mitochondrial genome contains 37 genes: 28 are encoded by the heavy strand, and nine by the light strand. Of the 37 genes, a total of 24 specify a mature RNA product: 22 mitochondrial tRNA molecules and two mitochondrial rRNA molecules, a 23S rRNA (a component of the large subunit of mitochondrial ribosomes) and a 16S rRNA (a component of the small subunit of the mitochondrial ribosomes). The remaining 13 genes encode polypeptides which are synthesized on mitochondrial ribosomes.

Each of the 13 polypeptides encoded by the mitochondrial genome is a subunit of one of the mitochondrial respiratory complexes, the multichain enzymes of oxidative phosphorylation which are engaged in the production of ATP. Note, however, that there are a total of about 100 different polypeptide subunits in the mitochondrial oxidative phosphorylation system, and so the vast majority are encoded by nuclear genes. All other mitochondrial proteins, including numerous enzymes, transport proteins, structural proteins etc., are encoded by the nuclear genome and are translated on cytoplasmic ribosomes before being imported into the mitochondria.

The mitochondrial genetic code: The mitochondrial genetic code (which is used to decipher only 13 different mitochondrial mRNAs on mitochondrial ribosomes) differs slightly from the nuclear genetic code (which specifies perhaps about 70 000–80 000 different mRNAs on cytoplasmic ribosomes). The mitochondrial genome encodes all the ribosomal RNA and tRNA molecules it needs for synthesizing proteins but relies on nuclear-encoded genes to provide all other components (such as the protein components of mitochondrial ribosomes, amino acyl tRNA synthetases, etc.).

As there are only 22 different types of human mitochondrial tRNA, individual tRNA molecules need to be able to interpret several different codons. Eight of the 22 tRNA molecules have anticodons which are each able to recognize families of four codons differing only at the third base, and 14 recognize pairs of codons which are identical at the first two base positions and share either a purine or a pyrimidine at the third base. Between them, therefore, the 22 mitochondrial tRNA molecules can recognize a total of 60 codons $[(8 \times 4) + (14 \times 2)]$. The remaining four codons, UAG, UAA, AGA and AGG cannot be recognized by mitochondrial tRNA and act as stop codons.

In addition to their differences in genetic capacity and different genetic codes, the mitochondrial and nuclear genomes differ in many other aspects of their organization and expression.

Coding and noncoding DNA: Unlike its nuclear counterpart, the human mitochondrial genome is extremely compact:

approximately 93% of the DNA sequence represents coding sequence. All 37 mitochondrial genes lack introns and they are tightly packed (on average one per 0.45 kb). The coding sequences of some genes (notably those encoding the sixth and eighth subunits of the mitochondrial ATPase) show some overlap and, in most other cases, the coding sequences of neighboring genes are contiguous or separated by one or two noncoding bases. Some genes even lack termination codons; to overcome this deficiency, UAA codons have to be introduced at the post-transcriptional level.

The only significant region lacking any known coding DNA is the displacement (D) loop region. This is the region in which a triple-stranded DNA structure is generated by synthesizing an additional short piece of the H-strand DNA, known as 7S DNA. The replication of both the H and L strands is unidirectional and starts at specific origins. In the former case, the origin is in the D loop and only after about two-thirds of the daughter H strand has been synthesized (by using the L strand as a template and displacing the old H strand) does the origin for L strand replication become exposed. Thereafter, replication of the L strand proceeds in the opposite direction, using the H strand as a template. The D loop also contains the predominant promoter for transcription of both the H and L strands. Unlike transcription of nuclear genes, in which individual genes are almost always transcribed separately using individual promoters, transcription of the mitochondrial DNA starts from the promoters in the D loop region and continues, in opposing directions for the two different strands, round the circle to generate large multigenic transcripts. The mature RNAs are subsequently generated by cleavage of the multigenic transcripts.

The nuclear genome is distributed between 24 different types of DNA duplex which show considerable regional variation in base composition and gene density

Size and banding patterns of human chromosomes: The nucleus of a human cell contains more than 99% of the cellular DNA. The nuclear genome is distributed between 24 different types of linear double-stranded DNA molecule, each of which has histones and other nonhistone proteins bound to it, constituting a chromosome. The 24 different chromosomes (22 types of autosome and two sex chromosomes, X and Y) can easily be differentiated by chromosome banding techniques, and have been classified into groups largely according to size and, to some extent, centromere position. In addition to the primary constriction (centromere) present on each chromosome, the long arms of chromosomes 1, 9 and 16 possess so-called secondary constrictions (light staining, apparently uncoiled chromosomal regions) which, like the centromeres, are composed of constitutive heterochromatin. By comparison with the size of a mitochondrial DNA molecule, an average size human chromosome has an enormous amount of DNA, approximately 130 Mb on average, but varying between approximately 50 and 260 Mb. In a 550 band metaphase chromosome preparation, an average band corresponds to about 6 Mb of DNA.

Base composition in the human nuclear genome: Since the entire nucleotide sequence of the human mitochondrial genome is known, its precise base composition is known. The sequence of the human nuclear genome is still being established (and is not expected to be finished before 2003), but current estimates suggest a figure of about 42% GC. However, the proportion of specific combinations of nucleotides can vary considerably. Like other vertebrate nuclear genomes, for example, the human nuclear genome has a conspicuous shortage of the dinucleotide CpG (that is, neighboring cytosine and guanine residues on the same DNA strand in the 5' → 3' direction). Taking the average figure of 42% GC, the individual base frequencies are : C = G = 0.21, and so the expected frequency for the dinucleotide CpG is $(0.21)^2 = 0.0441$. However, the observed frequency of the CpG dinucleotide is approximately one-fifth of this.

In vertebrate DNA, cytosine residues occurring in CpG dinucleotides are targets for methylation at carbon atom 5. Only about 3% of the cytosines in human DNA are methylated, but most that are methylated are found in the CpG dinucleotide, producing 5-methylcytosine. Over evolutionarily long periods of time, 5-methylcytosine spontaneously deaminates to give thymine and so CpG is continuously being depleted and replaced by TpG (or CpA on the complementary strand. Despite the overall background, certain small regions of DNA noted for their transcriptional activity are characterized by the expected CpG density.

Base composition can also show regional subchromosomal variation. For example, human telomeres are defined by numerous repeats of a 50% GC sequence, TTAGGG. Large tracts of condensed heterochromatin found at centromeric regions and in defined noncentromeric regions of many chromosomes are composed of specialized repetitive DNA whose sequence can vary considerably in base composition from the bulk DNA. Such variation is the basis for the ability of equilibrium density ultracentrifugation to fractionate the total DNA into subclasses. The alternating pale and light staining chromosome bands also differ in a number of features, with dark G bands being characterized by a lower GC content than the pale staining bands.

Gene density in the human nuclear genome: The total number of genes in the human genome has been estimated to be about 70 000–80 000. As all but 37 of these genes are located in the nuclear genome, this gives a rough estimate of about 3000 genes per chromosome. However, gene density can vary substantially between chromosomal regions and also between whole chromosomes. For example, heterochromatic regions are known to be very largely composed of repetitive noncoding DNA, and the centromeres and large regions of the Y chromosome, in particular, are notably devoid of genes.

Recently, insight into gene distribution along the lengths of the different chromosomes has been obtained by hybridizing purified CpG island fractions of the genome (which are associated with perhaps about 56% of human genes; Antequara and Bird, 1993) to metaphase chromosomes (Craig and Bickmore, 1994). On this basis, it is clear that gene density is high in subtelomeric regions and that some chromosomes (e.g. 19 and 22) are gene rich while others (e.g. 4 and 18) are gene poor. Differences between pale and dark G bands, which are, respectively, gene rich and gene poor, are illustrated by the contrast between the human leukocyte antigen (HLA) complex and the dystrophin gene (*DMD*) regions. The former is located in the pale G band, 6p21.3: at the time of writing, the most intensively investigated region of the HLA complex, the class III region, had 70 genes in 0.9 Mb of DNA, giving a density of one per 13 kb. By contrast, a full 2.4 Mb of DNA in the dark G band region, Xp21, appears to be devoted exclusively to the dystrophin gene.

Organization and distribution of human genes

The nuclear genome contains about 65 000–80 000 genes but only about 3% of the genome represents coding sequences

The number of genes in the human genome has been the subject of much speculation; while the small mitochondrial genome is known to have precisely 37 genes, the number in the nuclear genome remains unknown. Theoretical calculations based on the *mutational load* that a genome can tolerate and observed average mutation rates of human genes ($\sim 10^{-5}$ per gene per generation) suggest an upper limit of about 100 000. A variety of different approaches have been used to obtain more precise estimates of the total gene number. Three approaches have suggested a best estimate of about 65 000–80 000 genes:

- Genomic sequencing. Extrapolation from sequencing of large chromosomal regions may suggest that there are about 70 000 genes (Fields *et al.*, 1994). This is based on the observation that gene-rich regions have an average gene density of close to one per 20 kb, but gene-poor regions have a much lower density, say one-tenth of this density, and that the genome is split 50:50 into gene-rich and gene-poor regions.
- CpG island number. Restriction enzyme analysis using the methylation-sensitive enzyme *Hpa*II suggests that the total number of CpG islands in the human genome is 45 000 (Antequara and Bird, 1993). Using an estimate that approximately 56% of genes are associated with CpG islands, these authors have suggested a total of about 80 000 human genes.

- EST analysis. Large-scale random sequencing of cDNA clones provides so-called expressed sequence tags (ESTs, see Section 13.2.3). Comparison of known human EST sequences with a large set of different human genomic coding DNA sequences listed in sequence databases has suggested a figure of about 65 000 human genes (Fields *et al.*, 1994).

The above values suggest that the genes in the nuclear genome represent about 99.95% of the total number of cellular genes. If the average size of a human nuclear gene, including introns, is taken to be about 10–15 kb, this would mean that if the genes did not show overlaps, the total nuclear DNA occupied by genes would be about 70 000 × (10–15) kb or about 700–1050 Mb which corresponds roughly to about 25–35% of the genome. As the vast majority of nuclear genes encode polypeptides and the coding sequence required for an average size human polypeptide is taken to be about 500–600 codons, that is 1.5–1.8 kb, only about 3% of the nuclear genome (80–100 Mb of the 3300 Mb) would be expected to have a coding function.

RNA-encoding gene families often have numerous family members

While the great majority of human genes are expected to encode polypeptides, a significant minority encode mature RNA molecules of diverse function. The mitochondrial genome is exceptional in that 65% (24/37) of the genes encode RNA but even in the case of the nuclear genome about 5% of the genes, perhaps 3000–4000 genes in all, are expected to encode RNA molecules. In common with other cellular genomes a considerable variety of genes in the human genome are devoted to making mature RNA molecules which assist in the general process of gene expression. Some, notably rRNA and tRNA are involved in translation of mRNA. In addition, many other RNA families are involved in reactions leading to maturation not only of mRNA but also of rRNA, tRNA and other RNA species, involving both cleavage reactions and basespecific modification reactions. In addition, several other RNAs have diverse functions.

Ribosomal RNA (rRNA) genes: There are multiple rRNA genes. In addition to the two mitochondrial rRNA molecules, the 28S, 18S and 5.8S cytoplasmic rRNAs are encoded by a single transcription unit which is tandemly repeated about 250 times, comprising five clusters of about 50 tandem repeats located on the short arms of human chromosomes 13, 14, 15, 21 and 22. In addition, the 5S cytoplasmic rRNA is encoded by several hundred gene copies in at least three clusters on the long arm of chromosome 1. The major rationale for the repetition of cytoplasmic rRNA genes is likely to be based on gene dosage: by having a comparatively large number of these genes, the cell can satisfy the huge demand for cytoplasmic ribosomes needed for protein synthesis.

Transfer RNA (tRNA) genes: These belong to a very large dispersed gene family, comprising more than 40 different subfamilies each with several members which encode the different species of cytoplasmic tRNA. In addition to multiple copies of genes specifying the individual cytoplasmic tRNA molecules, there are several defective gene copies (*pseudogenes*).

Small nuclear RNA (snRNA) genes: A heterogeneous collection of several hundred small nuclear RNA species are encoded by a large dispersed family of genes. Many of the snRNA species are uridine-rich and are named accordingly, e.g. U3 snRNA means the third uridine-rich small nuclear RNA to be classified. Individual species of RNA are associated with specific proteins to form ribonucleoprotein particles (RNPs). Some are known to be important in RNA splicing. A large subfamily of perhaps about 200 genes are present in the nucleolus, and have been termed small nucleolar RNA (snoRNA). They have important roles in specific cleavage reactions and base-specific modifications during maturation of ribosomal RNA (see Smith and Steitz, 1997).

Other RNA genes: Additional RNA genes encode functionally diverse products, including the 7SL RNA component of the signal recognition particle which is required for protein export and the RNA component of telomerase, the enzyme required to synthesize DNA at the telomeres. More recently, evidence has been obtained suggesting that certain RNA genes encode products that are important in gene regulation. An important example is the *XIST* gene. This gene is thought to be the major gene involved in initiating the process of X chromosome inactivation, being expressed exclusively from inactivated X chromosomes. No long open reading frames can be identified, and gene function is thought to be carried out through an RNA product by a mechanism that remains obscure.

In addition, several RNA genes have been found at a variety of chromosomal regions that are known to be imprinted (imprinted genes are normally expressed from a maternally inherited copy or a paternally inherited copy, but not both; see Section 8.5.4). For example, the *H19* gene contains five exons and is expressed to give a polyadenylated cytoplasmic RNA which does not however associate with ribosomes. It shows a restricted pattern of expression during early development (fetal and neonatal liver, visceral endoderm and fetal gut) and is imprinted, since only the maternally inherited allele is expressed. Its functional significance is, however, unclear.

Functionally similar genes are occasionally clustered in the human genome, but are more often dispersed over different chromosomes

As seen in the previous section, some families of RNA genes are clustered. In the case of polypeptide-encoding gene families, some genes encoding identical or functionally related products are clustered, but often they are dispersed on several chromosomes.

Functionally identical genes: A very few human polypeptides are known to be encoded by two or more identical gene copies. Often, these are encoded by recently duplicated genes in a gene cluster, as in the case of the duplicated α-globin genes. In addition, some genes on different chromosomes encode identical polypeptides. Examples include members of histone gene subfamilies.

Functionally similar genes: A large fraction of human genes are members of gene families where individual genes are closely related but not identical in sequence. In many such cases the genes are clustered and have arisen by tandem gene duplication, as in the case of the different members of each of the α-globin and β-globin gene clusters. Genes which encode clearly related products but which are located on different chromosomes are generally less related, as in the case of the α-globin and β-globin genes. However, in the case of the *HOX* homeobox gene family which consists of clusters of approximately 10 genes on each of four chromosomes, individual genes on different chromosomes may be more related to each other than they are to members of the same gene cluster.

In addition to the above, genes encoding closely related tissue-specific isoforms, or subcellular compartment-specific isozymes are often nonsyntenic (i.e. located on different chromosomes).

Functionally related genes: Some genes encode products which may not be so closely related in structure, but are clearly functionally related. The products may be subunits of the same protein or macromolecular structure, components of the same metabolic or developmental pathway, or may be required to specifically bind to each other as in the case of ligands and their relevant receptors. In almost all such cases, the genes are not clustered and are usually found on different chromosomes.

Human genes show enormous variation in size and internal organization

Size diversity: Genes in simple organisms such as bacteria are comparatively similar in size, and usually very short. By contrast, complex organisms such as mammals show wide variation in gene size, a feature found especially in human genes which can vary in length from hundreds of nucleotides to several megabases. The enormous size of some human genes means that transcription can be time-consuming. For example, the human dystrophin gene requires about 16 hours to be transcribed, and transcripts undergo splicing before transcription is completed (Tennyson *et al.*, 1995).

As one would expect, there is a direct correlation between the size of a gene and the size of its product, but there are some striking anomalies. For example, apolipoprotein B has 4563 amino acids and is encoded by a 45 kb gene while the dystrophin gene is 2.4 Mb in length and encodes a product in muscle cells of 3685 amino acids.

Diversity in internal organization: There is an inverse correlation between gene size and the proportion of the gene length which is expressed at the RNA level. A very small minority of human genes lack introns and are generally very small genes. For those that do possess introns, the exon content as a percentage of gene length tends to be very small in large genes. This does not arise because exons in large genes are smaller than those in small genes: the average exon size in human genes is about 200 bp and, although very large exons are known, exon size is comparatively independent of gene length. Instead, the explanation is due to the huge variation in intron lengths: large genes tend to have very large introns. The relationship between gene and intron length is not, however, without anomalies: the human type 7 collagen gene (*COL7A1*) is an intermediate size gene (31 kb) but has a total of 118 exons and an average intron size of only 188 bp (Christiano *et al.*, 1994). The extraordinary number of exons in *COL7A1* is not even matched by the number of exons in the giant 2.4 Mb dystrophin (*DMD*) gene.

Rare examples of overlapping genes and genes within genes are known in the human genome

Partially overlapping genes: The genes of simple organisms are generally more clustered than those in complex organisms. The average gene density in the human genome is about one per 40–45 kb of DNA. Assuming a mean size of, say, 10–15 kb, human genes should be separated by about 30 kb of nongenic DNA on average. By contrast, average gene densities in simple organisms are very much higher: roughly one per 1, 2 and 5 kb, respectively, for *E. coli*, *Saccharomyces cerevisiae* and *Caenorhabditis elegans*. Simple genomes such as those of certain phages and bacteria often show examples of partially overlapping genes which use different reading frames, sometimes from a common sense strand. The human mitochondrial genome is another example of a simple genome packed with genetic information and it too has an example of such overlapping genes.

Genes within genes: The small nucleolar RNA (snoRNA) genes are unusual in that the majority of them are located within other genes, often ones which encode a ribosome-associated protein or a nucleolar protein. Possibly this arrangement has been maintained to permit coordinate production of protein and RNA components of the ribosome (Tycowski *et al.*, 1993).

In addition to the snoRNA genes there are a few examples of other genes being located within the introns of larger genes, and in some cases the internal genes as well as the host genes are known to encode polypeptides. Three illustrative examples are:

- *The neurofibromatosis type I (NF1) gene.* Intron 27 of the *NF1* gene spans about 40 kb and contains three small genes, each with two exons which are transcribed from the opposite strand to that used for the *NF1* gene.
- *The factor VIII gene.* Intron 22 of the blood clotting factor VIII gene (*F8C*) contains a CpG island from which two internal genes, *F8A* and *F8B* are transcribed in opposite directions. *F8A* is transcribed from the opposite strand to that used by the factor VIII gene. *F8B* is transcribed in the same direction as the factor VIII gene to give a short mRNA containing a new exon spliced on to exons 23–26 of the factor VIII gene.
- *The retinoblastoma susceptibility gene RB1.* Intron 17 of this gene is 72 kb long and contains a G protein-coupled receptor gene, U16, which is actively transcribed from the opposite strand.

Human multigene families and repetitive coding DNA

DNA sequences in the nuclear diploid genome usually exist as two allelic copies (on paternal and maternal homologous chromosomes). In addition to this degree of repetition, about 40% of the human nuclear genome in both haploid and diploid cells is composed of sets of closely related nonallelic DNA sequences (DNA sequence families or repetitive DNA). Within the considerable variety of different repetitive DNA sequences are DNA sequence families whose individual members include functional genes (multigene families), and also many examples of nongenic repetitive DNA sequence families.

The reassociation kinetics of human DNA suggest three broad classes of DNA sequence

Reassociation kinetics first suggested that complex genomes, such as the human genome, comprise different sequence classes on the basis of the copy number. Typically, human DNA is randomly sheared (e.g. by sonication) to give fragments whose average size is about 500 bp and the sheared DNA is denatured by heating to separate the complementary strands of each fragment.

Thereafter the DNA is cooled to a temperature of about 20–30°C below the melting temperature, T_m (which marks the mid-point of the transition between the double-stranded and single-stranded states of DNA heated in solution). The cooled DNA renatures but the rate of reassociation depends not only on time (t) but also on the initial concentration (C_0) of that sequence (i.e. the C_0t value).

The above type of analysis has suggested that the human genome consists of three broad sequence components:

- *Single copy*, or at least very low copy number, DNA (60%) - reassociates very slowly. A single strand from a single copy sequence will require some considerable time to find a complementary partner strand, given that the vast majority of DNA fragments are unrelated to it.
- *Moderately repetitive* (30%) - intermediate speed of reassociation.
- *Highly repetitive* (10%) - reassociates very rapidly. There are numerous copies of the same sequence and the chances of quickly finding complementary partners within the mass of different fragments are high.

Members of DNA sequence families can be identified by a variety of different approaches.

The operational definition of a DNA sequence family is the comparatively high level of DNA sequence similarity (sequence homology) between whole family members, or components of the family members. Members of a DNA sequence family can be identified and actively sought by a variety of methods:

- *DNA sequencing* - Allows direct calculations on the degree of sequence relatedness of family members.
- *DNA hybridization and cloning.* A probe from a gene family member typically gives a complex band pattern when hybridized against a Southern blot of genomic DNAs. Individual family members can then be cloned by screening genomic DNA libraries.
- *PCR cloning* - Permits identification of novel family members by designing degenerate primers corresponding to highly conserved nucleotide or amino acid sequences.

When two members of a repetitive DNA sequence family exhibit a high degree of sequence homology, a recent common evolutionary origin is indicated. As detailed in the following sections, DNA sequence families show considerable variation in the number of different repeat unit members in the family, the size of the repeating unit, chromosomal location, mode of repetition and capacity for expression.

Human gene families vary in the overall sequence relatedness of different family members and the extent to which particularly conserved subgenic sequences define the family

A large percentage of actively expressed human genes are members of families of DNA sequences which show a high degree of sequence similarity.

However, the extent of sequence sharing and the organization of family members can vary widely. Some family members may be nonfunctional (pseudogenes and gene fragments - see below) and rapidly accumulate sequence differences, leading to marked sequence divergence.

Classical gene families: Classical gene families are distinguished by members which exhibit a high degree of sequence homology over most of the gene length or, at least, the coding DNA component, a feature which automatically identifies such sequences as being closely related evolutionarily as well as functionally. In some cases, such as the individual rRNA gene families and individual histone gene families, there is an extremely high degree of sequence similarity between family members. Many other large gene families show a high degree of sequence similarity between family members.

Gene families encoding products with large, highly conserved domains: In some gene families there is particularly pronounced homology within specific strongly conserved regions of the genes; the corresponding sequence similarity between the remaining portion of the coding sequence in the different genes may be quite low. Often such families encode transcription factors that play important roles in early development, and the conserved sequence encodes a protein domain which is required to bind specifically to the DNA of selected target genes.

Gene families encoding products with very short conserved amino acid motifs: The members of some gene families may not be very obviously related at the DNA sequence level, but nevertheless encode gene products that are characterized by a common general function and the presence of very short conserved sequence motifs:

- the DEAD box gene family - contains several different genes whose products appear to function as RNA helicases and are characterized by the presence of eight short amino acid sequence motifs, including the DEAD box (the sequence Asp-Glu-Ala-Asp as represented by the one-letter amino acid code);
- the WD repeat gene family - gene products have a regulatory function, but there is considerable diversity in function. Typically characterized by tandem repeats with a central core of fixed length and containing small conserved amino acid motifs, including the WD (tryptophan-aspartate) sequence);
- the ankyrin repeat gene family - wide functional diversity but often involved in protein-protein interactions. Characterised by tandem repeats of a 33 amino acid sequence characterized by the presence of select amino acids at only a few positions;

- the LIM domain gene family - encode a characteristic cysteine-rich 56 amino acid domain most likely involved in protein-protein interactions.

Gene superfamilies: In some types of gene family, the genes encode products that are known to be functionally related in a general sense, and show only very weak sequence homology over a large segment, without very significant conserved amino acid motifs. Instead, there may be some evidence for general common structural features. Such genes, which appear to be evolutionarily related but more distantly than those in a classical or conserved domain/motif gene family, have been considered to be members of a gene superfamily. For example, in addition to the immunoglobulin gene family, other related genes such as the *HLA* genes, TCR genes, *T4* and *T8* genes are known to encode products with an immune system function and a domain structure that resembles that of immunoglobulins. Although, therefore, the level of sequence homology between such genes may be very low, the similarities in function and general domain structure have suggested the existence of a so-called Ig superfamily, in which there appears to be a distant common evolutionary relationship.

Human gene families can occur as closely clustered genes at specific subchromosomal locations, or as widely dispersed genes

A wide variety of human gene families have been identified and show considerable variation in both the organization of the genes and the extent to which individual genes within the gene family are related in terms of sequence and function. In terms of gene organization, two basic arrangements can be discerned: families where there is evidence of close gene clustering and families that are dispersed over several different chromosomal locations. This classification is rather arbitrary, however, since several gene families consist of multiple gene clusters at different chromosomal locations.

Gene families organized in a single cluster: Genes in an individual gene cluster are thought to arise by tandem gene duplication events. Different organizations are evident:

- Tandem gene organization. This arrangement is exemplified by the organisation of genes in individual rRNA gene clusters. The genes are highly related to each other in terms of both sequence and function, although certain family members may be nonfunctional.
- Close clustering. The individual genes may be more physically separate than in a tandemly repeated cluster but nevertheless may be closely clustered, even to the extent of being subject to a common regulatory mechanism (locus control region - see the example of the α- and β-globin gene clusters. Again the individual genes usually show a high degree of sequence and functional identity to each other, but many family members may be pseudogenes.
- Compound clusters. In other clustered gene families, however, the physical relationship between genes in a cluster may be less close and a cluster of related genes may also contain within it genes that are unrelated in sequence and function, a compound gene cluster. For example, the HLA complex on 6p21.3 is dominated by families of genes which encode class I and II HLA antigens and various serum complement factors, but individual family members may be separated by functionally unrelated genes such as members of the steroid 21-hydroxylase gene family, etc. The gene organization in the latter case shows evidence of an ancestral duplication event that involved a segment containing different genes that are unrelated in function and sequence.

Gene families organized in multiple gene clusters

Some gene families are organized in clusters distributed over two or more chromosomal locations. Again different organizations are evident, with some families showing very high similarity between genes on different clusters, while others are marked by comparatively low sequence homology between genes on different clusters.

- High cluster similarity. Here sequence homology between gene family members on different chromosomes may be very high as in the case of the different rRNA gene clusters on the short arms of chromosomes 13, 14, 15, 21 and 22. Another useful example is the olfactory receptor gene family which encodes a diverse repertoire of receptors which allow us to discriminate thousands of different odors. The family consists of perhaps 1000 genes, making it the largest in the human genome, and is organized in large clusters at more than 25 different chromosomal locations.
- Low cluster similarity. Often sequence homology is greater within a cluster than between clusters. For example, the globin gene family includes genes at three locations: the α-globin cluster on 16p, the β-globin cluster on 11p and the myoglobin gene at 22q and although all globin genes are clearly related, those within a cluster are more related to each other than they are to genes in one of the other clusters.

Interspersed gene families: Some gene families show no obvious physical relationship between family members, which are dispersed as solitary genes at two or more different chromosomal locations. The family members may show considerable sequence divergence unless their dispersion has been a relatively recent event, or there has been considerable selection pressure to maintain sequence conservation. The following examples illustrate some of the many different types of organization.

- *Families expected to originate from two genomes.* Current ideas on mitochondrial origins have envisaged that the original genome was derived from a prokaryote but that many of the genes were subsequently transferred to the nuclear genome. Some families of nuclear genes encode cytoplasm-specific and mitochondrial-specific isoforms for certain enzymes and other key metabolic products (see *Table 7.5* for some examples).
- *Families originating from ancient genome duplication or gene duplication events.* This class of interspersed

gene families typically contains only a few members, as in the case of the *PAX* gene family, and appear to have evolved by a combination of gene duplication and/or genome duplication events over a long period of evolutionary time. Typically, all or most of the genes are functional, and may individually encode highly related products.

- *Families originating largely by retrotransposition events.* Some gene families have expanded comparatively recently in evolutionary terms by a process whereby RNA transcribed from one or a small number of functional genes is converted by cellular reverse transcriptase into natural cDNA which then becomes integrated elsewhere in the chromosomes. Most such copies are nonfunctional.

Pseudogenes, truncated gene copies and gene fragments are commonly found in multigene families

Families of RNA genes or polypeptide-encoding genes are frequently characterized by defective copies of essentially all of the gene or its coding sequence (pseudogenes), or a portion of it, in some cases a single exon (gene fragments). A large variety of different classes are found. The following examples are meant to be illustrative of the types of defective gene copies found in different types of gene family.

Nonprocessed pseudogenes in a gene cluster: Individual gene clusters are typically characterized by the presence of defective gene copies which have been copied *at the level of genomic DNA*. This means that the defective gene copies can contain copies of the exons, introns and promoter regions of the functional genes (nonprocessed pseudogenes). Classical examples are the α-globin and β-globin clusters. Only one member of the α-globin gene family is a nonprocessed pseudogene, but three of the seven gene copies in the β-globin gene cluster are. Another β-globin gene family member, *HBQ1*, is likely to be an expressed pseudogene: it encodes a type of globin polypeptide, θ-globin, but there is no evidence that the latter is ever incorporated into a haemoglobin molecule and so θ-globin is likely to lack any function.

Truncated genes and gene fragments in a gene cluster: The class I HLA gene family at 6p21.3 is a classical example of a gene cluster which is characterized by nonprocessed pseudogenes, truncated gene copies and gene fragments. Although the number of class I *HLA* genes can vary on different chromosome 6s, comprehensive analysis of one of these identified 17 family members clustered over about 2 Mb. Six of the genes are known to be expressed, although the precise functions of some of these are still not clearly understood.

The remaining members are clearly defective. Four are conventional full-length pseudogenes, but another five represent truncated gene copies (lacking the 5' end in four cases and the 3' end in the other case) and two are fragments which contain a small component of the gene sequence, even a single exon.

Nonprocessed pseudogenes in an interspersed gene family: For several polypeptide-encoding genes, defective copies containing intronic sequence have been found elsewhere in the genome. Two illustrative examples are:

- *The* NF1 *(neurofibromatosis type I) gene.* This gene is located close to the chromosome 17 centromere at 17q11.2 and has at least 11 nonprocessed pseudogene or gene fragment copies, nine of which are located at pericentromeric regions on seven different chromosomes (Regnier *et al.,* 1997). Characterization of the chromosome 15-specific *NF1* gene copies revealed that they contain copies of a segment of the *NF1* gene spanning exons 8 and 27 and including the intron sequences. This type of gene duplication may be a result of what has been called pericentromeric plasticity.
- *The* PKD1 *(adult polycystic kidney disease) gene.* This gene consists of 46 exons spanning 50 kb at 16p13.3. A truncated 5' gene copy comprising approximately 70% of the gene and encompassing exons 1 to 34 and all the intervening introns has been faithfully replicated at least three times and inserted into a more proximal location at 16p13.1 (the European Polycystic Kidney Disease Consortium, 1994).

An apparently related case concerns the *CFTR* (cystic fibrosis) gene at 7q31.2. which has 24 exons spanning 250 kb. Copies of a ~30 kb sequence encompassing exon 9 and flanking intron sequences are distributed at a large number of chromosomal locations. However, in this case a likely explanation proposed by Rozmahel *et al.* 1997 is a retrotransposition model. Transcription from the sense strand of the *CFTR* gene was envisaged to generate an antisense transcript encompassing the exon 9 sequence and flanking intron regions. The transcript could then have been converted to cDNA by cellular reverse transcriptase prior to integration at several sites in the genome.

Processed pseudogenes in an interspersed polypeptide-encoding gene family: Interspersed gene families frequently have several defective gene copies which have been copied *at the cDNA level.* This form of retrotransposition is carried out by cellular reverse transcriptases, which transcribe mRNA into natural cDNA which can then integrate into chromosomal DNA at sites of temporary breakage. Although processed pseudogenes are typically not expressed, several examples are known where natural cDNA copies of a gene appear to be expressed, often in a testis-specific fashion. Often the functional intron-containing gene locus in such families is located on the X chromosome.

Processed pseudogenes in an RNA-encoding gene family: Although the size of some interspersed polypeptide-encoding gene families testifies to the success of retrotransposition as a mechanism for generating processed gene copies, the really successful (in terms of high copy number) retrotranspositions have been performed from RNA polymerase III transcripts. For example, the Alu repeat family is considered to have arisen as processed pseudogenes copied from the 7SL RNA gene. Genes such as this which are transcribed by RNA polymerase III often contain an internal promoter which facilitates the expression of newly transposed copies.

The coding sequences of many human genes contain repeated sequence motifs

In addition to the types of gene sequence duplication discussed in the previous sections, many human genes, like other eukaryotic genes, contain intragenic repeated sequences. Repeated sequences in coding DNA may involve different forms of repeated structure, including comparatively large sequences encoding protein domains or small sequence motifs. Different modes of repetition can be seen but tandem repetition is rather common.

- Tandem repetition of microsatellite sequences is common and may simply reflect statistical expected frequencies for certain base compositions. For example, certain genes which have a very high % GC often have long runs of single cytosines or guanines in the coding sequence. Such sequences are comparatively unstable and are prone to single nucleotide deletion or insertion events which are often pathogenic. A variety of genes are known to have long tracts of certain trinucleotides in their coding sequences and runs of the CAG trinucleotides in the coding sequences of several genes have been found to be capable of expansion causing pathogenesis.
- Tandem repetition of sequences encoding known or assumed protein domains is quite common, and may be functionally advantageous in some cases by providing a more available biological target. In some cases, the sequence homology between the repeats can be very high; in other cases it may be rather low.

Extragenic repeated DNA sequences and transposable elements

The human nuclear genome, like that of other complex eukaryotes, contains a large amount of highly repeated DNA sequence families which are largely transcriptionally inactive. A wide variety of different repeats are known and dedicated repeat sequence databases have been established. Like multigene families, noncoding repetitive DNA shows two major types of organization: tandemly repeated and interspersed.

Tandemly repeated noncoding DNA: Such families are defined by blocks (or arrays) of tandemly repeated DNA sequences. Individual arrays can occur at a few or many different chromosomal locations. Depending on the average size of the arrays of repeat units, highly repetitive noncoding DNA belonging to this class can be grouped into three subclasses: satellite, minisatellite and microsatellite DNA.

In addition to these three major subclasses (which are detailed in the following three sections), a fourth class has recently been recognized and described as megasatellite or macrosatellite DNA. Despite the name, this type of DNA is characterized by array lengths which can be comparatively modest compared to some satellite DNA arrays. Instead, the prefix mega- has been used to emphasize the large size of the repeating unit which can be several kilobases long. This class is exemplified by the RS447 megasatellite which consists of about 60 tandem copies of a novel 4.7 kb repeat on 4p15, plus another array of several copies on distal 8p. Array lengths can be highly polymorphic and the RS447 repeat is reported to contain a putative open reading frame of 1590 bp.

Interspersed repetitive noncoding DNA: The individual repeat units are not clustered, but are dispersed at numerous locations in the genome, and together account for perhaps one third of the DNA in the human genome (Smit, 1996). Most of the DNA families belonging to this class contain some members that are capable of undergoing retrotransposition (i.e. transposition through an RNA intermediate).

The chromosomal locations of different types of tandemly repeated DNA can show a very restricted or highly dispersed pattern, whereas different classes of interspersed repeat DNA can show preferential location within different types of chromosome bands.

Satellite DNA is composed of very long arrays of tandem repeats which can be separated from bulk DNA by buqyant density gradient centrifugation

Human satellite DNA is comprised of very large arrays of tandemly repeated DNA with the repeat unit being a simple or moderately complex sequence. Repeated DNA of this type is not transcribed and accounts for the bulk of the heterochromatic regions of the genome, being notably found in the vicinity of the centromeres (pericentromeric heterochromatin). The base composition, and therefore density, of such DNA regions is dictated by the base composition of their constituent short repeat units and may diverge substantially from the overall base composition of bulk cellular DNA.

Isolation by buoyant density gradient centrifugation: When DNA is isolated from human cells by conventional methods, it is subject to mechanical shearing. Fragments are generated from the bulk DNA (with a base composition of ~42% GC) and fragments from the satellite DNA regions which may have a similar or different base composition. If the base composition is significantly different, satellite DNA sequences can be separated from the bulk DNA by buoyant density gradient centrifugation. Following centrifugation, they appear as minor (or satellite) bands of different buoyant density from a major band which represents bulk DNA. Typically, human DNA is complexed with Ag^+ ions and then fractionated in buoyant density gradients containing cesium sulfate, whereupon three satellite bands are identified at different densities: satellite 1 – 1.687 gcm^{-3}; satellite 2 – 1.693 gcm^{-3}; satellite 3 – 1.697 gcm^{-3}.

Each of these satellite classes includes a number of different tandemly repeated DNA sequence families (satellite subfamilies), some of which are shared between different classes. DNA sequence analysis has revealed that some of the repetitive DNA families in the satellites are based on very simple repeat units. For example, both satellite 2 and satellite 3 contain sequence arrays which are based on tandem repetition of the sequence ATTCC. Additionally, restriction mapping has revealed satellite subfamilies which show additional higher order repeat units superimposed on the small basic repeat units. Such subfamilies are thought to arise as a result of subsequent amplification of a unit which is larger than the initial basic repeat unit and contains some diverged units.

Alphoid DNA and centromeric heterochromatin: Other types of satellite DNA sequence cannot easily be resolved by density gradient centrifugation. They were first identified by digestion of genomic DNA with a restriction endonuclease which typically has a single recognition site in the basic repeat unit. In addition to the basic repeat unit size (monomer), such enzymes will produce a characteristic pattern of multimers of the unit length because of occasional random loss of the restriction site in some of the repeats (Singer, 1982a). Alpha satellite (or alphoid DNA) constitutes the bulk of the centromeric heterochromatin and accounts for about 3–5% of the DNA of each chromosome. It is characterized by tandem repeats of a basic mean length of 171 bp, although higher order units are also seen. The sequence divergence between individual members of the alphoid DNA family can be so high that it is possible to isolate chromosome-specific subfamilies for each of the human chromosomes (Choo *et al.*, 1991).

The function of satellite DNA remains unclear. The centromeric DNA of human chromosomes largely consists of various families of satellite DNA. Of these, only the α-satellite is known to be present on all chromosomes, and its repeat units often contain a binding site for a specific centromere protein, CENP-B. Recently cloned α-satellite arrays have been shown to seed *de novo* centromeres in human cells, indicating that α-satellite plays an important role in centromere function.

Minisatellite DNA is composed of moderately sized arrays of tandem repeats and is often located at or close to telomeres

Minisatellite DNA comprises a collection of moderately sized arrays of tandemly repeated DNA sequences which are dispersed over considerable portions of the nuclear genome. Like satellite DNA sequences, they are not normally transcribed (but see below).

Hypervariable minisatellite DNA: Hypervariable minisatellite DNA sequences are highly polymorphic and are organized in over 1000 arrays (0.1–20 kb long) of short tandem repeats (Jeffreys, 1987). The repeat units in different hypervariable arrays vary considerably in size, but share a common core sequence, GGGCAGGAXG (where X = any nucleotide), which is similar in size and in G content to the *chi* sequence, a signal for generalized recombination in *E. coli*. While many of the arrays are found near the telomeres, several hypervariable minisatellite DNA sequences occur at other chromosomal locations.

The great majority of hypervariable minisatellite DNA sequences are not transcribed, except for elements occuring within noncoding intragenic sequences. Some, however, are expressed. For example, the *MUC1* locus on 1q is an expressed hypervariable minisatellite locus. It encodes a glycoprotein found in several epithelial tissues and body fluids which is highly polymorphic as a result of extensive variation in the number of minisatellite-encoded repeats (Swallow *et al.*, 1987).

The significance of hypervariable minisatellite DNA is not clear, although it has been reported to be a 'hotspot' for homologous recombination in human cells (Wahls *et al.*, 1990). Nevertheless it has found many applications. Various individual loci have been characterized and used as genetic markers, although the preferential localization in subtelomeric regions has limited their use for genome-wide linkage studies. A major application has been in DNA fingerprinting, in which a single DNA probe which contains the common core sequence can hybridize simultaneously to multiple minisatellite DNA loci on all chromosomes, resulting in a complex individual-specific hybridization pattern.

Telomeric DNA: Another major family of minisatellite DNA sequences is found at the termini of chromosomes, the telomeres. The principal constituent of telomeric DNA is 10–15 kb of tandem hexanucleotide repeat units, especially TTAGGG, which are added by a specialized enzyme, telomerase. By acting as buffers to protect the ends of the chromosomes from degradation and loss and by providing a mechanism for replicating the ends of the linear DNA of chromosomes, these simple repeats are directly responsible for telomere function.

Microsatellite DNA is defined by the presence of short arrays of tandem simple repeat units and is dispersed throughout the human genome

Microsatellite DNA families include small arrays of tandem repeats which are simple in sequence (often 1–4 bp) and are interspersed throughout the genome. Of the mononucleotide repeats, runs of A and of T are very common and together account for about 10 Mb, or 0.3% of the nuclear genome. By contrast, runs of G and of C are very much rarer. In the case of dinucleotide repeats, arrays of CA repeats (TG repeats on the complementary strand) are very common, accounting for 0.5% of the genome, and are often highly polymorphic. CT/AG repeats are also common, occurring on average once every 50 kb and accounting for 0.2% of the genome, but CG/GC repeats are very rare.

This is so because C residues which are flanked at their 3' end by a G residue (i.e. CpG) are prone to methylation and subsequent deamination, resulting in TpG (or CpA on the opposite strand). Trinucleotide and tetranucleotide tandem repeats are comparatively rare, but are often highly polymorphic and increasingly have been investigated to develop highly polymorphic markers.

The significance of microsatellite DNA is not known. Alternating purine-pyrimidine repeats, such as tandem repeats of the dinucleotide pair CA/TG, are capable of adopting an altered DNA conformation, *Z*-DNA, *in vitro*, but there is little evidence that they do so in the cell. Although microsatellite DNA has generally been identified in intergenic DNA or within the introns of genes, a few examples have been recorded within the coding sequences of genes. Tandem repeats of three nucleotides in coding DNA may be sites that are prone to pathogenic expansions.

Highly repeated interspersed DNA families contain a small percentage of actively transposing DNA elements

Two major classes of mammalian interspersed repetitive DNA families have been discerned on the basis of repeat unit length (Singer, 1982b): SINEs and LINEs.

SINEs (short interspersed nuclear elements): The most conspicuous human SINE is the *Alu* repeat family (so called because of early attempts at characterizing the sequence using the restriction nuclease *Alu*I). The *Alu* repeat contains an internal RNA polymerase III promoter sequence. It has attained a very high copy number in the human genome and appears to have originated by retrotransposition from the 7SL RNA gene. The *Alu* repeat is primate-specific but other mammals have similar types of sequence derived from the 7SL RNA gene such as the B1 family in mouse.

Unlike the *Alu* repeat, another major human SINE family is not restricted to primates, with copies being found in marsupials and monotremes. In accordance with its distribution this family has been termed the MIR (mammalian-wide interspersed repeat) family.

LINEs (long interspersed nuclear elements): Human LINEs are exemplified by the *LINE-1* or *L1* element (also called the *Kpn* repeat because of early attempts at characterizing this family using the restriction nuclease *Kpn*I). The LINE-1 element is also found in other mammals such as the mouse.

In addition to the human *Alu* and LINE-1 repeat families, there are many smaller families, including the THE-1 (transposable human element family), many MER (medium reiteration frequency) families and families of human endogenous retroviruses (HERV) or retrovirus-like elements (RTLV).

Some members of the interspersed repeat families have been considered as transposable elements, unstable DNA elements which can migrate to different regions of the genome. Rare examples are known of human DNA sequences which appear to have been copied by a DNA-mediated transposition event (transposons) and a variety of sequences have been found in the human genome which resemble known DNA transposons such as the *mariner* transposon and others (Smit and Riggs, 1996). However, the latter are not thought to be actively transposing. Instead, the great majority of human transposable elements undergo retrotransposition, that is their RNA transcripts can be converted within the cell to a complementary DNA form which can reinsert back into chromosomal DNA at a variety of different locations. Three classes of mammalian sequence are known to be able to transpose through an RNA intermediate. In each case, the transposition event involves duplication of a very short sequence at the target site, causing the transposed sequence to be flanked by short repeats.

The Alu repeat occurs about once every 3 kb in the human genome and includes examples that are transcribed

The *Alu* repeat is the most abundant sequence in the human genome, with a copy number of about 1 000 000 (see Deininger, 1989; Smit, 1996). *Alu* repeats have a relatively high GC content and, although dispersed mainly throughout the euchromatic regions of the genome, have been reported to be preferentially located in R chromosome bands (Korenberg and Rykowski, 1988). The latter correspond to the pale bands seen when using standard Giemsa staining and represent the most transcriptionally active regions of the genome. The full-length *Alu* repeat is about 280 bp long and is usually flanked by short (often 6–18 bp) *direct repeats* (i.e. the repeats are in the same orientation). The typical *Alu* sequence is a tandemly repeated dimer, with the repeats sharing an approximately 120 bp sequence followed by a short sequence which is rich in A residues on one strand and T residues on the complementary strand. However, there is asymmetry between the tandem repeats: one repeat unit contains an internal 3'-bp sequence lacking in the other. Monomers, containing only one of the two tandem repeats, and various truncated versions of dimers and monomers are also common.

The two repeated units of the *Alu* sequence show a striking resemblance to the sequence for 7SL RNA, a component of the signal recognition particle, which facilitates transport of proteins across the membrane of the endoplasmic reticulum. Because of this and the observation of the A-rich regions and the flanking direct repeats, it has been widely assumed that the *Alu* sequence has been propagated by retrotransposition from 7SL RNA, and therefore represents a processed 7SL RNA pseudogene. Certainly, transposition by *Alu* sequences is known to occur [presumably as a result of *trans*-acting cellular reverse transcriptases such as those encoded by LINE-1 (*Kpn*) elements, and may occasionally cause clinical problems]. Possibly, the very high copy number achieved by this processed pseudogene is related to the presence of a promoter sequence in the 7SL RNA sequence (the 7SL RNA gene, like tRNA genes is transcribed by RNA polymerase III from an internal promoter).

By contrast, processed pseudogenes from RNA polymerase II transcripts lack promoter sequences and their only chance of expression is if the integration event places them next to a functional promoter sequence.

Currently, the function of the *Alu* sequence, if any, is unknown, although several roles have been considered (see Schmid, 1998). Although the average expected frequency is one copy per 3 kb, high density clustering of *Alu* repeats is known to occur in certain regions. Because of their ubiquity, *Alu* sequences have been considered to promote unequal recombination, a mechanism which while occasionally causing disease, may be evolutionarily advantageous in promoting gene duplication. Although conspicuously absent from coding sequences, *Alu* sequences are often found in noncoding intragenic locations, notably in introns and occasionally in untranslated sequences. Consequently, they are often represented in the primary transcript RNA from genes encoding polypeptides, and occasionally in mRNA. The *Alu* sequence can also be transcribed from its internal promoter by RNA polymerase III *in vitro*, and *in vivo* transcription of some *Alu* sequences can result in the accumulation of a small cytoplasmic RNA which can be specifically bound by two cytoplasmic signal recognition particle proteins, SRP9 and SRP14 (Chang *et al.*, 1994).

The LINE-1 (Kpn repeat) includes examples that appear to encode a reverse transcriptase and are actively transposing

The human LINE-1 (L1, *Kpn* repeat) family is expected to consist of over 100 000 and perhaps as many as 400 000– 500 000 interspersed repeats (Smit, 1996). Of these, multiple members are known to be actively transposing (Sassaman *et al.*, 1997). The full-length consensus element is 6.1 kb long and has two open reading frames (ORFs). ORF1 is located close to one end (conventionally known as the 5' end) of the consensus element and encodes a protein of unknown function, p40 (so called because its molecular weight is ~40 kDa). ORF2 encodes a protein with an endonuclease domain and also a reverse transcriptase domain (see Sassaman *et al.*, 1997 ; Kazazian and Moran, 1998). The full-length consensus element contains an internal promoter within a region of untranslated DNA preceding ORF1 (conventionally called the 5' -UTR) while at the other end there is an $(A)_n/(T)_n$ sequence, often described as the 3' poly(A) tail. As in the case of other elements that transpose, the LINE-1 elements are flanked by short duplicated repeats.

The full-length LINE-1 element is comparatively rare and most repeats are truncated at the 5' end, resulting in a population that is heterogeneous in length but sharing a common 3' end with the poly(A) tail. LINE-1 elements are primarily located in euchromatic regions but show an inverse relationship with *Alu* repeats by appearing to be located preferentially in the dark G bands (Giemsa positive) of metaphase chromosomes. Like the *Alu* repeats, they are conspicuously absent from coding sequences but may be found in intragenic noncoding sequences. As a result, they may be represented in the primary RNA transcript of large genes, but they are conspicuously absent from coding sequences.

5

HUMAN GENE EXPRESSION

AN OVERVIEW OF GENE EXPRESSION IN HUMAN CELLS

The control mechanisms used to regulate human gene expression are fundamentally similar to those found in other mammals, and generally resemble those in eukaryotes in general. Although much more complex than equivalent mechanisms in organisms with small genomes, many of the same basic principles apply and as in other eukaryotes, a major level at which gene expression is controlled is the initiation of transcription. Mammals are particularly complex multicellular organisms and so it is perhaps unsurprising that there are some gene control mechanisms which are not used in bacteria or in some other eukaryotes. Various regulation mechanisms are required to maintain many different facets of mammalian gene expression, both at the spatial and temporal levels. Although simplistic, it is convenient to consider three broad levels at which gene regulation can operate.

Transcriptional regulation of gene expression: We have long been accustomed to the idea that a primary control of gene regulation in eukaryotes occurs at the level of initiation of transcription. Regulation of expression can occur through the core promoter of a gene, at the level of recruitment and processivity of the relevant RNA polymerase. Expression of genes is initiated by the binding of transcription factors to the promoter. Basal levels of transcription can be modulated by binding of protein factors to other regulatory regions occurring in the sequences flanking the gene or sometimes within introns of the gene.

Post-transcriptional regulation of gene expression: This category overlaps with the previous section since it includes mechanisms operating at the level of RNA processing, such as RNA splicing which may more accurately be considered as co-transcriptional rather than posttranscriptional (Steinmetz, 1997). In addition to RNA processing, other levels at which control of gene expression can be exerted include: mRNA transport, translation, mRNA stability, protein processing, protein targeting, protein stability, etc.

A surprising variety of mechanisms are employed at the level of RNA processing with single alleles in an individual often able to generate a variety of different gene products (isoforms). The occurrence of these and other mechanisms has required a more flexible definition of the term *gene* than has formerly been used. Several mechanisms are involved in regulating gene expression at the level of translation and an increasing number of regulatory sequences have been identified in the 5' and especially 3' untranslated regions of mRNA. Control of gene expression at the level of protein processing, targeting and stability has been shown in certain systems. For example, activation of some peptide hormones such as insulin requires post-translational cleavage from precursor forms.

Epigenetic mechanisms and long range control of gene expression: In addition to genetic factors, additional factors which can be transmitted to progeny cells following cell division *but which are not directly attributable to the DNA sequence* are described as epigenetic. DNA methylation is an epigenetic mechanism which plays an important part in mammalian gene control, acting as a general method of maintaining repression of transcription.

In addition, a variety of other mechanisms which affect the chromatin environment of a gene and hence its capacity for gene expression are known to operate in mammalian cells. In some cases, the mechanisms ensure that within a cell only one of the two parentally inherited alleles is normally expressed, even although the nucleotide sequence of the allele which is not expressed may be identical to the one which is expressed.

Control of gene expression by binding of *trans*-acting protein factors to *cis*-acting regulatory sequences in DNA and RNA

A common molecular basis for much of the control of gene expression (whether it occurs at the level of initiation of transcription, RNA processing, translation or RNA transport) is the binding of protein factors to regulatory nucleic acid sequences. The latter can be DNA sequences found in the vicinity of the gene or even within it, or RNA transcript sequences at the level of precursor RNA or mRNA. As the protein factors engaged in regulating gene expression are themselves encoded by distantly located genes, they are required to migrate to their site of action, and so are called *trans*-acting factors. In contrast, the regulatory sequences to which they bind to are on the same DNA or RNA molecule as the gene or RNA transcript that is being regulated. Such sequences are said to be *cis*-acting.

Control by DNA-protein binding: A major control of gene expression in eukaryotic cells is exerted at the level of *initiation* of transcription where three different types of RNA polymerase are known to transcribe different classes of genes. All three types of RNA polymerase are large enzymes, consisting of 8–14 subunits, and in each case the polymerase is recruited to transcribe a gene following binding of proteins (transcription factors) to specific regulatory DNA sequences within the gene or in its vicinity. Chromatin is a highly organized and densely packed structure which does not easily afford access to RNA polymerases and so transcription factors are required to help activate it to give a more open structure that will enable transcription to take place.

Control by RNA-protein binding: In addition to transcription factors, RNA-binding proteins are used to regulate gene expression. The best-studied examples involve binding to regulatory sequences in the untranslated sequences of mRNA, permitting translational control of gene expression. In addition, specific RNA-protein binding interactions are expected to be involved in the control of gene expression at the level of differential RNA processing too, as in the case of binding of SR and HnRNP proteins to pre-mRNA in order to modulate the choice of exons in splicing.

Ubiquitous transcription factors are required for transcription of RNA polymerases I and III

RNA polymerases I and III in eukaryotic cells are dedicated to transcribing genes to give RNA molecules (rRNA, tRNA, etc.) which assist in expression of the polypeptide-encoding genes. The transcribed genes are housekeeping genes since rRNA and tRNA are required in essentially all cells to assist in protein synthesis. As a result, ubiquitous transcription factors are required to assist RNA polymerases I and III.

Transcription by RNA polymerase I: RNA polymerase I is confined to the nucleolus and is devoted to transcription of the 18S, 5.8S and 28S rRNA genes. The latter are consecutively organized on a common 13 kb transcription unit. A compound unit of the 13-kb transcription unit and an adjacent 27 kb nontranscribed spacer is tandemly repeated about 50–60 times on the short arms of each of the five human acrocentric chromosomes, at the nucleolar organizer regions. The resulting five clusters of rRNA genes, each about 2 Mb long, are referred to as ribosomal DNA or rDNA.

Initiation of transcription of the 28S, 5.8S and 18S rRNA genes is initiated following binding of two transcription factors to a core promoter element at the transcription initiation site and an upstream control element located over 100 nucleotides upstream. One of the transcription factors, UBF (upstream binding factor), is a homodimer and its identical subunits may bind first to the core promoter element and upstream control element, bringing them together so that they can be bound by the second factor, SL1. The bound transcription factors subsequently recruit RNA polymerase I to form an initiation complex.

The primary transcript expressed from the single 13 kb transcription unit is a 45S precursor rRNA which undergoes a variety of cleavage reactions and base-specific modifications (carried out by a large number of different types of small nucleolar RNA (snoRNA)) to generate the mature 28S, 5.8S and 18S rRNA species. Thus, these genes differ from the vast majority of nuclear genes, which are individually transcribed. Instead, rDNA transcription resembles mtDNA transcription: both result in multigenic transcripts which yield functionally related products. This unusual use of polygenic primary transcripts is no different in principle, however, from the way in which a single primary translation product is occasionally cleaved to generate two or more functionally related polypeptides.

Transcription by RNA polymerase III: RNA polymerase III is also involved in transcription of a variety of housekeeping genes, encoding various small stable RNA molecules such as 5S rRNA, tRNA molecules, 7SL RNA and some of the snRNA molecules needed for RNA splicing. These genes are characterized by promoters that lie *within* the coding sequence of the gene, rather than upstream of it. In tRNA genes, the promoter is bipartite, consisting of two well conserved sequences, the A box and the B box, while in the 5S rRNA gene, a single promoter element is present, the C box. In each case, transcription by RNA polymerase III is thought to proceed by binding of ubiquitous transcription factors to the promoter elements, followed by subsequent binding of other factors and finally recruitment of the polymerase.

Transcription of polypeptide-encoding genes often requires complex sets of cis-acting transcriptional control sequences and tissue-specific transcription factors

RNA polymerase II is responsible for transcribing all genes which encode polypeptides and also certain species of snRNA gene. Like RNA polymerases I and III, RNA polymerase II is dependent on auxiliary general transcription factors (the usual nomenclature has a common prefix TF to denote transcription factor followed by a Roman numeral to denote the associated RNA polymerase). In the case of RNA polymerase II, there are a variety of auxiliary transcription factors such as TFIIA, TFIIB, TFIID, TFIIE, TFIIF, TFIIH, etc, which can be complex in structure (Nikolov and Burley, 1997). For example, the TATA box-binding protein (TBP), is only one of the multiple protein subunits that make up TFIID and the associated proteins are known as TBP-associated factors, or TAF proteins. The complex of polymerase and general transcription factors is known as the basal transcription apparatus; it constitutes all that is required to initiate transcription. Genes are constitutively expressed at a minimum rate determined by the core promoter unless the rate of transcription is increased or switched off by additional positive or negative regulatory elements (which may be located some distance away *or by intrinsic components of the promoter itself*).

Some of the genes which encode polypeptides are housekeeping genes, but unlike the products of genes transcribed by RNA polymerases I and III, a large percentage of genes transcribed by RNA polymerase II show tissue-restricted or tissue-specific expression patterns. Since the DNA in different nucleated cells of an individual is essentially identical, the identity of a cell, whether it be a hepatocyte or a T lymphocyte for instance, is defined by the proteins made by the cell. In addition to general ubiquitous transcription factors, therefore, tissue-specific or tissue-restricted transcription factors regulate the expression of many genes which encode polypeptides, by recognizing and binding specific *cis*-acting sequence elements.

Partly because of the large size of mammalian nuclear genomes and also because of the general need for more sophisticated control systems imposed by having very large numbers of interacting genes, control elements in eukaryotic cells are quite elaborate. Often, regulation of expression of individual human genes is controlled by several sets of *cis*-acting regulatory elements. While the individual regulatory elements may be composed of multiple short sequence elements (typically 4–8 nucleotides long) distributed over a few hundred base pairs, the different classes of regulatory element which modulate the expression of a single gene may be located at considerable distances. A variety of different types of *cis*-acting elements can be recognized, including promoters, enhancers, silencers, boundary elements (insulators) and response elements.

Tissue specificity and developmental stage specificity of gene expression is often conferred by enhancer and silencer sequences and a variety of *cis*-acting sequences have been identified which are specifically recognized by tissue-specific transcription factors. For example, specific expression in erythroid cells is often signalled by one of two sequences: TGACTCAG (or its reverse complement CTGAGTCA) which are recognized by the erythroid-specific transcription factor NF-E2, or by the sequence (A/T)GATA(A/G) or its reverse complement which are recognized by the GATA series of erythroid specific transcription factors.

In addition to actively promoting tissue-specific transcription, some *cis*-acting silencer elements confer tissue or developmental stage specificity by blocking expression in all but the desired tissue. For example, the neural restrictive silencer element (NRSE) represses expression of several genes in all tissues other than neural tissues (Schoenherr *et al.*, 1996). A transcription factor that binds to the NRSE and which is variously called the neural restrictive silencer factor (NRSF) or the RE-1 silencing transcription factor (REST) is ubiquitously expressed in non-neural tissue and neuronal precursors during early development but subsequently it is specifically *not expressed* in more mature (postmitotic) neurons.

Transcription factors contain conserved structural motifs that permit DNA binding

Transcription factors recognize and bind a short nucleotide sequence, usually as a result of extensive complementarity between the surface of the protein and surface features of the double helix in the region of binding. Although the individual interactions between the amino acids and nucleotides are weak (usually hydrogen bonds, ionic bonds and hydrophobic interactions), the region of DNA-protein binding is typically characterized by about 20 such contacts, which collectively ensure that the binding is strong and specific. In human and other eukaryotic transcription factors, two distinct functions can often be identified and located in different parts of the protein:

- An activation domain. As the name suggests, this type of domain functions in activating transcription of the target genes once the transcription factor has bound to it. Activation domains are thought to stimulate transcription by interacting with basal transcription factors so as to assist the formation of the transcription complex on the promoter. Although not so well-studied as DNA-binding domains, some are known to be rich in aspartate and glutamate residues (*acidic activation domains*); others are rich in proline or glutamate.

- A DNA-binding domain. This type of domain is necessary to permit specific binding of the transcription factor to its target genes. In contrast to activation domains, DNA-binding domains of transcription factors have been well-studied. A number of conserved structural motifs have been identified which are common to many different transcription factors with quite different specificities, including the leucine zipper, helix-loop-helix, helix-turn-helix, and zinc finger motifs which are described below. Each of the motifs uses α-helices (or occasionally β-sheets) to bind to the major groove of DNA. Clearly, although the motifs in general provide the basis for DNA binding, the precise collection of sequence elements in the DNA-binding domain will provide the basis for the required sequence-specific recognition. Most transcription factors bind to DNA as homodimers, with the DNA-binding region of the protein usually distinct from the region responsible for forming dimers.

The leucine zipper motif: The leucine zipper is a helical stretch of amino acids rich in leucine residues (typically occurring once every seven amino acid residues, i.e. once every two turns of the helix), which readily forms a dimer. Each monomer unit consists of an amphipathic *a-helix* (hydrophobic side groups of the constituent amino acids face one way; polar groups face the other way). The two α-helices of the individual monomers join together over a short distance to form a coiled-coil with the predominant interactions occurring between opposed hydrophobic amino acids of the individual monomers. Beyond this region the two α-helices separate, so that the overall dimer is a Y-shaped structure. The dimer is thought to grip the double helix much like a clothes peg grips a clothes line. In addition to forming homodimers, leucine zipper proteins can occasionally form heterodimers depending on the compatibility of the hydrophobic surfaces of the two different monomers. Such heterodimer formation provides an important combinatorial control mechanism in gene regulation.

The helix-loop-helix motif: The helix-loop-helix (HLH) motif is related to the leucine zipper and should be distinguished from the helix- turn-helix (HTH) motif described in the next section. It consists of two α-helices, one short and one long, connected by a flexible loop. Unlike the short turn in the HTH motif, the loop in the HLH motif is flexible enough to permit folding back so that the two helices can pack against each other; that is, the two helices lie in planes that are parallel to each other, in contrast to the two helices in the HTH motif. The HLH motif mediates both DNA binding and protein dimer formation and it permits occasional heterodimer formation. In the latter case, however, heterodimers form between a full-length HLH protein and a truncated HLH protein which lacks the full length of the α-helix necessary to bind to the DNA. The resulting heterodimer is unable to bind DNA tightly. As a result, HLH dimers are thought to act as a control mechanism, by enabling inactivation of specific gene regulatory proteins.

The helix-turn-helix motif: The HTH motif is a common motif found in homeoboxes, and a number of other transcription factors. It consists of two short α-helices separated by a short amino acid sequence which induces a turn, so that the two α-helices are orientated differently (i.e. the two helices do not lie in the same plane, unlike those in the HLH motif). The structure is very similar to the DNA-binding motif of several bacteriophage regulatory proteins such as the λ cro protein whose binding to DNA has been intensively studied by X-ray crystallography. In the case of both the λ cro protein and eukaryotic HTH motifs, it is thought that while the HTH motif in general mediates DNA binding, the more C-terminal helix acts as a specific recognition helix because it fits into the major groove of the DNA, controlling the precise DNA sequence which is recognized.

The zinc finger motif: The zinc finger motif involves binding a zinc ion by four conserved amino acids to form a loop (finger), a structure which is often tandemly repeated. Although several different forms exist, common forms involve binding of a Zn^{2+} ion by two conserved cysteine residues and two conserved histidine residues, or by four conserved cysteine residues. The resulting structure may then consist of an α-helix and a β-sheet held together by coordination with the Zn^{2+} ion, or of two α-helices. In either case, the primary contact with the DNA is made by an α-helix binding to the major groove. The so-called Cys_2/His_2 finger typically comprises about 23 amino acids with neighboring fingers separated by a stretch of about seven or eight amino acids.

A variety of mechanisms permit transcriptional regulation of gene expression in response to external stimuli

In eukaryotic cells, gene expression can be altered in a semipermanent way as cells differentiate, or in a temporary, easily reversible way in response to extracellular signals (inducible gene expression). Environmental cues such as the extracellular concentrations of certain ions and small nutrient molecules, temperature, shock, etc., can result in dramatic alteration of gene expression patterns in cells exposed to changes in these parameters. In complex multicellular animals there are also fundamental requirements for cells to communicate with each other and different modes of cell signaling are possible. In some cases, alteration of gene expression is conducted at the translational level which can offer certain advantages. In other cases, gene expression is altered by modulating transcription.

Transcriptional regulation in response to cell signaling can take different forms, but the endpoint is always the same: a previously inactive transcription factor is specifically activated by the signaling pathway and then subsequently binds to specific regulatory sequences located in the promoters of target genes, thereby activating their transcription. In the case of transcription regulated by signaling molecules or their intermediaries, such regulatory sequences are often referred to as response elements.

Ligand-inducible transcription factors: Small hydrophobic hormones and morphogens such as steroid hormones, thyroxine and retinoic acid are able to diffuse through the plasma membrane of the target cell and bind intracellular receptors in the cytoplasm or nucleus. These receptors (often called hormone nuclear receptors) are inducible transcription factors. Following binding of the homologous ligand, the receptor protein associates with a specific DNA response element located in the promoter regions of perhaps 50–100 target genes and activates their transcription.

Although thyroxine and retinoic acid are structurally and biosynthetically unrelated to the steroid hormones, their receptors belong to a common nuclear receptor superfamily. Two conserved domains characterize the family: a centrally located DNA binding domain of about 68 amino acids, and a ~240 amino acid ligand-binding domain located close to the C terminus. The DNA binding domain contains zinc fingers and binds as a dimer with each monomer recognizing one of two hexanucleotides in the response element. The two hexanucleotides are either inverted repeats or direct repeats which are typically separated by three or five nucleotides. In the absence of the ligand, the receptor is inactivated by direct repression of the DNA binding domain function by the ligand-binding domain, or by binding to an inhibitory protein, as in the case of the glucocorticoid receptor.

Activation of transcription factors by signal transduction: Unlike lipid-soluble hormones or morphogens, hydrophilic signaling molecules such as polypeptide hormones, cannot diffuse through the plasma membrane. Instead, they bind to a specific receptor on the cell surface. After binding of the ligand molecule, the receptor undergoes a conformational change and becomes activated in such a way that it passes on the signal via other molecules within the cell (signal transduction). Various classes of cell surface receptor are known but many of them have a kinase activity or can activate intracellular kinases. Signal transduction pathways are often characterized by complex regulatory interplay between kinases and phosphatases which can activate or repress intermediates by phosphorylation/dephosphorylation. In many cases, the phosphorylation or dephosphorylation induces an altered conformation. In the case of activation of a signaling molecule, the altered conformation often means that a signaling factor is no longer inhibited by some repressor sequence present in an inhibitory protein to which it is bound, or in a domain or sequence motif within its own structure.

In terms of transcriptional activation, two general mechanisms permit rapid transmission of signals from cell-surface receptors to the nucleus, both involving protein phosphorylation:

- protein kinases are activated and then translocated from the cytoplasm to the nucleus where they phosphorylate target transcription factors;
- inactive transcription factors stored in the cytoplasm are activated by phosphorylation and translocated into the nucleus.

The following two sections provide examples to illustrate these two mechanisms (see also Karin and Hunter, 1995).

Hormonal signaling through the cyclic AMP pathway

Cyclic AMP is an important second messenger which acts in response to a variety of hormones and other signaling molecules. It is synthesized from ATP by a membrane bound enzyme, adenylate cyclase. Hormones which activate adenylate cyclase bind to a cell surface receptor which is of the G protein-coupled receptor class. Binding of the hormone to the receptor promotes the interaction of the receptor with a G protein which consists of three subunits, α, β and γ. Following this interaction the α subunit of the G protein is activated, causing it to dissociate and stimulate adenylate cyclase.

The increase in intracellular cAMP produced by activated adenylate cyclase can then activate the transcription of specific target sequences that contain a cAMP response element or CRE. This function of cAMP is mediated by the enzyme protein kinase A. Cyclic AMP binds to protein kinase and activates it by permitting release of the two catalytically active subunits which then enter the nucleus and phosphorylate a specific transcription factor, CREB (CRE-binding protein). Activated CREB then activates transcription of genes with the cAMP response element.

Activation of NF-êB via protein kinase C signaling: NF-êB is a transcription factor which is involved in a variety of aspects of the immune response. In its inactive state, NF-êB is retained in the cytoplasm where it is complexed with an inhibitory subunit, IêB. However, the latter can be targeted for degradation following phosphorylation by protein kinase C. The consequent destruction of IêB permits NF-êB to translocate to the nucleus and activate its various target genes. Protein kinase C is activated by diacylglycerol. The latter is produced when binding of various growth factors and hormones to specific cell surface receptors triggers activation of receptor-linked phospholipase C activity. The activated enzyme converts PIP_2 (phosphatidylinositol 4,5 bisphosphate) to IP_3 (inositol 1,4,5-trisphosphate) and diacylglycerol.

Translational control of gene expression can involve specific recognition by RNA-binding proteins of regulatory sequences within the untranslated sequences of mRNA

Different forms of translational control of gene expression are evident and an increasing number of eukaryotic and mammalian mRNA species have been shown to contain regulatory sequences in their untranslated sequences (most frequently at the 3' end; see Wickens *et al.*, 1997; Day and Tuite, 1998). Several eukaryotic and mammalian RNA-binding proteins have also been identified and shown to bind to specific regulatory sequences present in untranslated sequences, thereby providing the basis for translational control of gene expression (Siomi and Dreyfuss, 1997). A variety of different RNA-binding domains have been identified and they include elements which have previously been associated with DNA-binding properties of transcription factors such as zinc fingers and homeodomains.

Intracellular RNA localization: The interaction between *cis*-acting regulatory elements in RNA and *trans*-acting RNA-binding proteins can be envisaged to alter RNA structure in various ways: facilitating or hindering interactions with other *trans*-acting factors; altering higher order RNA structure; bringing together initially remote RNA sequences; or providing localization or targeting signals for transport of RNA molecules to specific intracellular locations. In the latter case, numerous eukaryotic and mammalian mRNAs are known to be transported as RNP particles to specific locations within cells, and transport on microtubules and actin filaments has been demonstrated in some cases to require specific molecular motors (Hazelrigg, 1998). For example, tau mRNA is localized to the proximal portions of axons rather than to dendrites, where many mRNA molecules are located in mature neurons, and myelin basic protein mRNA is transported with the aid of kinesin to the processes of oligodendrocytes.

A rationale for such intracellular mRNA localization mechanisms is that they may provide a more efficient way to localize protein products than protein targeting: as detailed below, a single mRNA can give rise to many different protein molecules, assuming that it can engage with ribosomes. Different sequential steps have been envisaged: initial translational repression, transport within the cell, localization (to the specific subcellular destination) and then localization-dependent translation. Recently, key regulatory sequences which are required for various steps in this process have been identified in the untranslated sequences, predominantly the 3' UTR, of many mRNA species (Hazelrigg, 1998). For example, two elements within the 3' UTR of myelin basic protein mRNA are required for distinct steps in its transport to the processes of oligodendrocytes: a 21 nucleotide RNA transport sequence and a longer RNA localization region (Ainger *et al.,* 1997).

Translational control of gene expression in response to external stimuli: Translational control of gene expression can permit a more rapid response to altered environmental stimuli than the alternative of activating transcription. Iron metabolism provides two useful examples. Increased iron levels stimulate the synthesis of the iron-binding protein, ferritin, without any corresponding increase in the amount of ferritin mRNA. Conversely, decreased iron levels stimulate the production of transferrin receptor (TfR) without any effect on the production of transferrin receptor mRNA. The 5' -UTR of both ferritin heavy chain mRNA and light chain mRNA contain a single iron-response element (IRE), a specific *cis*-acting regulatory sequence which forms a hairpin structure. Several such IRE sequences are also found in the 3' UTR of the transferrin receptor mRNA (see Klausner *et al.*, 1993). Regulation is exerted by binding of IREs by a specific IRE-binding protein which is activated at low iron levels.

Translational control of gene expression during early development: Gene expression during oocyte maturation and at the earliest embryonic stages is regulated at the level of translation, not transcription. Following fertilization of a human oocyte, no mRNA is made initially until the 4–8 cell stage when zygotic transcription is activated, that is, transcription of the genes present in the zygote. Before this time, cell functions are specified by maternal mRNA that was previously synthesized during oogenesis. While it is presently unclear to what extent the regulation of human gene expression parallels that of model organisms at this stage, extrapolation from the latter would suggest that a variety of mRNAs are stored in oocytes in an inactive form, characterized by having short oligo (A) tails.

Such mRNAs were previously subject to deadenylation and the resulting short oligo (A) tail means that they cannot be translated. Subsequently, at fertilization or later in development, the stored inactive mRNA species can be activated by cytoplasmic polyadenylation, restoring the normal size poly (A) tail. Cytoplasmic polyadenylation appears to use the same type of poly (A) polymerase activity as in the standard polyadenylation of newly formed mRNA (which occurs in the nucleus). However, in addition to the AAUAAA signal, the mRNA needs to have a uridine-rich upstream *cytoplasmic polyadenylation element* (Wahle and Kuhn, 1997). Another mechanism that is used to regulate translation of some mRNAs during development is translational repression (masking) whereby RNA-binding proteins can recognize and bind specific sequences in the 3' UTRs of the mRNAs, thereby repressing translation.

Alternative transcription and processing of individual genes

In addition to the control that is exerted in selecting specific genes (or their transcripts) for activation or repression, control mechanisms can also select between specific alternative transcripts of a single gene. Differential promoter usage or differential RNA processing events can result in a large number of different isoforms and these and other mechanisms have challenged the classical definition of a gene.

Transcription of a single human gene can be initiated from a variety of alternative promoters and can result in a variety of tissue-specific isoforms

Several human and mammalian genes are known to have two or more alternative promoters, which can result in different isoforms with different properties (see Ayoubi and van de Ven, 1996). The isoforms can provide:

- tissue-specificity (a frequent occurrence; see the example of the human dystrophin gene below);
- developmental stage-specificity (e.g. the insulin-like growth factor II gene);
- differential subcellular localization (e.g. soluble and membrane-bound isoforms);
- differential functional capacity (as in the case of the progesterone receptor);
- sex-specific gene regulation.

One of the most celebrated examples of differential promoter usage in humans concerns the giant dystrophin (*DMD)* gene which comprises a total of more than 79 exons distributed over about 2.4 Mb of DNA in Xp21. At least eight different alternative promoters can be used (Cox and Kunkel, 1997).

Four of the alternative promoters are located near the conventional start site and comprise a brain cortex-specific promoter, a muscle-specific promoter located 100 kb downstream, a promoter which is used in Purkinje cells of the cerebellum and located a further 100 kb downstream, and a lymphocyte-specific promoter.

Usage of these promoters results in large isoforms with a molecular weight of 427 kDa (referred to as Dp427 where Dp = Dystrophin protein and often given a suffix to indicate tissue specificity e.g. Dp427-M to indicate the muscle specific isoform).

The four Dp427 isoforms differ in their extreme N-terminal amino acid sequence as a result of using four different alternatives for exon 1.

In addition to the four alternative promoters encoding the conventional large isoforms, at least four other alternative internal promoters can be used. Transcription from these promoters uses only a downstream subset of the exons, resulting in significantly smaller isoforms: a Dp260 isoform produced in retinal cells; a Dp140 isoform produced by many cells in the brain and kidney; a Dp116 isoform produced in Schwann cells and a small Dp71 isoform produced in many cell types. Note that the alternative usage of promoters enforces alternative use of exons but that alternative splicing events which are independent of differential promoter usage are also very common. In the case of dystrophin, for example, additional isoform complexity is introduced by alternative splicing, especially at the C terminus.

Human genes often encode more than one product as a result of alternative splicing and alternative polyadenylation events

In addition to differential use of promoters which can enforce alternative use of exons, a variety of alternative RNA processing events can also result in alternative isoforms. The primary mechanisms are alternative splicing events (that is, distinct from those induced by differential promoter usage), and alternative polyadenylation events. In many cases a combination of these mechanisms can result during the processing of a single gene. Together with the additional possibility of differential promoter usage, these mechanisms can result in very large numbers of isoforms for a single gene.

Alternative splicing: A large percentage of human genes undergo alternative splicing whereby different exon combinations are included in transcripts from the same gene during RNA processing. For many genes, numerous isoforms can be generated at the RNA level, but often the functional significance is poorly appreciated. In some genes, alternative splicing results in very considerable diversity in the untranslated regions. For example, in the liver alternative splicing results in at least 8 different 5' UTR sequences for human growth hormone receptor mRNA (Pekhletsky *et al.,* 1992), but the functional significance, if any, is not understood.

Alternative splicing of coding sequence exons is also common and some of the resulting protein isoforms have been shown to be tissue specific, so that individual exons present in one isoform but not in others may be termed 'muscle-specific' , 'brain-specific' etc. The different isoforms can provide a variety of possibilities for altered functional properties but detailed knowledge of the functional significance of the different isoforms is still comparatively sparse.

The best understood model system for understanding the regulation of splicing is the sex determination pathway in *Drosophila* which also controls gene dosage. Alternative splicing is used in each branch of this pathway to control the expression of transcriptional regulators or chromatin-associated proteins that influence transcription, and both positive and negative control of splicing is evident (Lopez, 1998). In mammalian cells candidate splice regulators are the SR family of RNA-binding proteins (which have a distinctive C terminal domain rich in serine (S)-arginine (R) dipeptides] and some HnRNP (heterogeneous nuclear ribonucleoprotein particle) proteins. These proteins are known to promote various steps in assembly of spliceosomes and they are also known to bind to splicing enhancer sequences, regulatory sequences which can enhance splice site recognition.

Alternative polyadenylation: The usage of alternative polyadenylation signals is also quite common in human mRNA, and different types of alternative polyadenylation have been identified (see Edwards-Gilbert *et al.,* 1997). In many genes, two or more polyadenylation signals are found in the 3' UTR and the alternatively polyadenylated transcripts can show tissue specificity; in other cases, alternative polyadenylation signals may be brought into play following alternative splicing. As an example of the latter, a combination of alternative splicing and alternative polyadenylation of the calcitonin gene (*CALC*) results in tissue-specific expression of two isoforms. Calcitonin, a circulating Ca^{2+} homeostatic hormone, is produced in the thyroid;

the calcitonin gene-related peptide (CGRP), which may have both neuromodulatory and trophic activities, is synthesized in the hypothalamus.

RNA editing is a rare form of post-transcriptional processing whereby base-specific changes are enzymatically introduced at the RNA level

RNA editing is a form of post-transcriptional processing which can involve enzyme-mediated insertion or deletion of nucleotides or substitution of single nucleotides *at the RNA level*. Insertion or deletion RNA editing appears to be a peculiar property of gene expression in mitochondria of kinetoplastid protozoa and slime molds. Substitution RNA editing is frequently employed in some systems, such as the mitochondria and chloroplasts of vascular plants where individual mRNAs may undergo multiple C → U or U → C editing events, and has also been observed in a few mammalian genes (Ashkenas, 1997). At least four different classes of RNA editing are known to occur in human cells:

- C⇒U editing. Human APOB lipoprotein mRNA editing has been well-studied. In the liver the APOB gene encodes a 14.1 kb mRNA transcript and a 4536 amino acid product, apoB100. However, in the intestine the same gene encodes a 7 kb mRNA which contains a premature stop codon not present in the gene and encodes a product, apoB48, which is identical in sequence to the first 2152 amino acids of apoB100. A specific cytosine deaminase, Apobec1, converts a single cytosine at nucleotide 6666 in the intestinal APOB mRNA to uridine, thereby generating a stop codon.
- A⇒I editing. Genes encoding some ligand-gated ion channels including glutamate receptors and related proteins are subject to this type of mRNA editing. An adenosine is deaminated to give inosine (I), a base not normally present in mRNA (the amino group at carbon 6 of adenosine is replaced by a C=O carbonyl group). Inosine base pairs preferentially with cytosine and also interacts with ribosomes during translation as if it were a G. In the case of the glutamate receptor B gene, for

example, RNA editing replaces a CAG (glutamine) codon by CIG which is translated as if it were the CGG codon (arginine). This type of editing which brings about a Gln⇒Arg is often referred to as Q/R editing after the single letter code for the two amino acids involved.

- Other classes of RNA editing. Two other documented forms of editing in human mRNA are the U⇒C editing in mRNA from the WT1 Wilms' tumor gene and U⇒A editing in α-galactosidase mRNA (Ashkenas, 1997).

Asymmetry as a means of establishing differential gene expression and DNA methylation as means of perpetuating differential expression

The concept of tissue specificity of human gene expression is long established. What is much less clear is how such patterns get laid down initially. Since the DNA content of all nucleated cells in an organism is virtually identical, genetic mechanisms cannot explain how differential gene expression first develops in cells. To explain this, CH Waddington evoked *epigenetic* mechanisms of gene control during development. In recent times, a variety of epigenetic mechanisms have been identified, including ones which can perpetuate particular states of gene expression in somatic cell lineages.

Selective gene expression in cells of mammalian embryos most likely develops in response to short range cell-cell signaling events

In order to explain subsequent tissue-, cell- and developmental stage-specific patterns of expression, some mechanism is required to set up an asymmetry or axis in the fertilized egg cell or in very early development. In *Drosophila*, the egg is inherently asymmetrical because of transfer of gene products from asymmetrically sited nurse cells. The embryo develops initially as a multinucleate syncytium (effectively one big cell) and regionalization depends on the response of individual nuclei to long-range gradients of regulatory molecules. In mammals, however, the egg cell is relatively small and early embryonic development creates an apparently symmetrical aggregate of individual cells. Nevertheless, development becomes asymmetric. The generation of asymmetry in mammalian cells could derive from early positional clues. Some aspects of early development are inherently asymmetrical including the point of entry of the sperm during fertilization, the attachment of the embryo to the uterine wall during implantation and the location of cells with respect to their neighbors. As the embryo develops into a ball of cells, and later on as more complex structures develop, individual cells will vary in the number of cell neighbors available. Short range intercellular signaling events (by direct cell-cell signaling or short-range intercellular signaling events) can provide a means of identifying cell position, and triggering differential gene expression. For example, if an intercellular signaling molecule has a range of, say, one cell diameter, then the cells at the outside of the blastula will receive different signals from those surrounded by neighbors on all sides, and the different positional cues may be translated into differential gene expression. As particular cell systems develop during, for example, organogenesis (mostly accomplished between the 4th and 9th embryonic weeks), particular cell type growth or differentiation factors may then induce the expression of developmental stage-specific and/or tissue-specific transcription factors.

Vertebrate DNA methylation is very largely confined to CpG dinucleotides and patterns of DNA methylation can be inherited when cells divide

Once differential expression patterns have been set up, epigenetic mechanisms can ensure that differential expression patterns are stably inherited when cells divide. DNA methylation is thought to play a major role in this respect, permitting the stable transmission from a diploid cell to daughter cells of chromatin states which repress gene expression. However, the precise function of DNA methylation in eukaryotes is still imperfectly understood and clearly shows species differences. Some organisms, for example, have no detectable DNA methlyation, as in the case of *Drosophila, C. elegans* and the yeast *Saccharomyces cerevisiae.* In those organisms where DNA methylation does occur, the patterns and functions of DNA methylation may differ.

Patterns of DNA methylation in vertebrates: The pattern of vertebrate DNA methylation differs from that in bacterial cells. In the latter, adenine and cytosine can both be methylated but in vertebrates methylation of DNA is restricted to cytosine residues. Only about 3% of the cytosines in human DNA are methylated, but most that are methylated are found in the CpG dinucleotide (that is, the methylated cytosines are almost always ones whose 3' carbon atom is linked by a phosphodiester bond to the 5' carbon atom of a guanine). In addition, a much smaller percentage of methylated cytosines occur within the sequence CpNpG.

Cytosine residues occurring in CpG dinucleotides in vertebrate DNA are targets for methylation by a specific cytosine methyltransferase. Methylation occurs at carbon atom 5 of the cytosine to generate 5-methylcytosine, which is chemically unstable and can spontaneously deaminate to give thymine. Over long periods of evolutionary time, the number of CpGs in vertebrate DNA has gradually been eroded, although regions of the normal (expected) CpG frequency are known and often mark transcriptionally active sequences.

Maintenance and de novo methylation during development: Unlike bacterial methylases, vertebrate cytosine methyltransferases show a strong preference for recognizing a *hemi-methylated* DNA target (i.e. one that is already methylated on one strand only). The sequence CpG shows dyad symmetry and so, following DNA replication, the newly synthesized DNA strands will receive the same CpG methylation pattern as the parental DNA. As a result, the CpG methylation pattern can be stably transmitted to daughter cells. The perpetuation of a pre-existing methylation pattern is sometimes known as maintenance methylation and is carried out in mammalian cells by the product of the *Dnmt1* gene.

The pattern of 5-methlycytosine distribution in the genome of differentiated somatic cells varies according to cell type but maintenance methylation ensures that methylation patterns in individual somatic cell lineages are quite stable. During gametogenesis and in the developing embryo, however, there are dramatic changes in methylation. The genomes of the primordial germ cells of the embryo are not methylated to any extent. After gonadal differentiation and as the germ cells begin to develop, *de novo* methylation occurs leading to substantial methylation of the DNA of mammalian sperm and egg cells. The sperm genome is more heavily methylated than the egg's genome, and *sex-specific differences in methylation patterns* are found, notably at imprinted loci. The *Dnmt1* methyltransferase gene, in addition to being the predominant maintenance DNA methyltransferase in mammalian cells may also be the major *de novo* methyltransferase. It is highly expressed in male germ cells, mature oocytes and in the early embryo. *Dnmt1* gene expression has been shown to be subject to sex-specific regulation with oocyte- and spermatocyte-specific promoters introducing oocyte- and spermatocyte-specific exons leading to different gene products.

The genome of the fertilized oocyte is an aggregate of the sperm and egg genomes and so it and the very early embryo are substantially methylated with methylation differences at paternal and maternal alleles of many genes. Later on, at the morula and early blastula stages in the preimplantation embryo, genome-wide demethylation occurs. Later still, at the pregastrulation stage, there is widespread *de novo* methylation. However, the extent of this methylation varies in different cell lineages:

- *the somatic cell lineage* is heavily methylated;
- *trophoblast-derived lineages* which give rise to the placenta, yolk sac, etc., are undermethylated;
- *early primordial germ cells* are spared; their genomic DNA remains very largely unmethylated until after gonadal differentiation and as the germ cells develop whereupon widespread *de novo* methylation occurs.

DNA methylation in animals has been thought to act as a form of host defense against transposons as well as a way of perpetuating patterns of transcriptional repression

Although not all eukaryotes appear to be subject to DNA methylation, its function in animal cells does appear to be critically important, and targeted mutagenesis of the cytosine methyltransferase gene in mice results in embryonic lethality. The precise function of DNA methylation in animal cells, however, remains unclear. Current views have focused in particular on two aspects of animal cells: the genome size (animals have comparatively large genomes with large numbers of genes, and also large numbers of highly repetitive DNA families belonging to the transposon class); and the mode of development (especially the variation in terms of lifespan and rate of cell turnover). Two quite contrasting views regarding the primary function of DNA methylation in animal cells have been the subject of much controversy: the host defense model and the gene regulation model.

Host defense as a primary function for DNA methylation: Like the restriction-modification function of DNA methylation in bacteria, the host-defense model envisages that the primary function of DNA methylation in animal cells is to confer a form of genome protection, but in this case checking the spread of transposons. About one-third of the DNA sequence in the human genome can be classified as belonging to (retro)transposon families and a small fraction of these sequences in the human genome and other genomes is known to be actively transposing. Transposon families in the human and other genomes are known to be heavily methylated (about 90% of the 5-methylcytosines are thought to be located in retrotransposon families) and so DNA methylation has been viewed as a mechanism for repressing such transposition, which if left unchecked could be expected to be damaging to cells. However, recently obtained data from an invertebrate chordate, *Ciona intestinalis*, appear to be inconsistent with the genome defense model: multiple copies of an apparently active retrotransposon and a large fraction of highly repeated SINEs were predominantly unmethylated, while genes, by contrast, appeared to be methylated (Simmen *et al.*, 1999).

Gene regulation as the primary function for DNA methylation: DNA methylation in vertebrates has been viewed as a mechanism for silencing transcription and may constitute a default position. DNA sequences which are transcriptionally active require to be unmethylated (at least at the promoter regions). While DNA methylation in invertebrates may serve to repress transposons and other repeated sequence families, it may have acquired a special role in vertebrates as a mechanism for regulating expression of endogenous genes and reducing transcriptional noise (by silencing a large fraction of genes whose activity is not required in a cell). By reducing unnecessary gene expression, DNA methylation may have permitted the increase in gene number and in complexity that characterizes vertebrates (Bird, 1995). The counterargument is that the methylation status of the 5' regions of tissue-specific genes cannot be correlated with expression in different tissues, and that the role of methylation in gene expression is in specialized biological functions resulting from mechanisms (e.g. imprinting) which use allele-specific gene expression.

DNA methylation and gene expression: The DNA of transcriptionally active and inactive chromatin differs in a number of features including the degree of compaction and the extent of its methylation. While methylation of CpG islands downstream of promoters does not block continued transcription through these regions, there is no doubt that methylated promoter regions are correlated with transcriptional silencing. In addition, the extent of histone acetylation is an important factor. Specific histone acetyltransferases add acetyl groups to lysine residues close to the N terminus of histone proteins. The acetylated N termini then form tails that protrude from the nucleosome core. As the acetylated histones are thought to have a reduced affinity for the DNA, and possibly for each other, the chromatin may be able to adopt a more open structure that is more suited to gene expression. Deacetylation of the histones, however, promotes repression of gene expression presumably because the chromatin can become more condensed.

Recently, the processes of DNA methylation and histone deacetylation have been shown to be linked. Repression at methlyated CpG sequences in promoter regions appears to be mediated by proteins which specifically bind to methylated CpG. Two of these proteins have been identified, MeCP1 and MeCP2 (methylated CpG binding proteins 1 and 2), and the latter has been shown to be essential for embryonic development and to function as a transcriptional repressor. The ability of MeCP2 to repress gene expression has been shown to involve a histone deacetylase complex. One possible model envisages that an initial signal for transcriptional repression is the binding of MeCP2 to methylated CpG promoter sequences. The bound MeCP2 protein is then recognized by a complex consisting of a transcriptional repressor and a histone deacetylase which removes the acetyl groups from the N termini of the histones, so that the chromatin becomes more condensed.

Long-range control of gene expression and imprinting

Chromatin structure may exert long-range control over gene expression

A predominant theme in eukaryotic gene expression, distinguishing it from bacterial gene expression, has been that genes are *individually* transcribed. Promoters and related upstream elements typically control expression of a single gene with a transcription start point located within 1 kb of the control element. Some *cis*-acting elements, however, exert long-range control over a much larger chromosomal region and there is increasing evidence for coordinate regulation of gene clusters.

Studies where genes are repositioned elsewhere in the genome have also suggested that chromosomes are organized into functional domains of gene expression (chromatin domains). For example, when genes are translocated to new chromosomal regions (either as a result of spontaneous chromosome breakage events, or by genetically manipulating model organisms - see Section 21.3), aberrant gene expression may often occur, even although the entire gene and the required control sequences in its immediate flanking sequences are preserved intact. Neighboring chromosome domains are envisaged to be separated by boundary elements (also called insulators) which act as barriers to the effects of distal enhancers and silencers (Geyer, 1997).

Competition for enhancers or silencers: Sometimes long-range control of gene expression appears to depend on competition between clustered genes for an enhancer.

Heterochromatin-induced position effects: Studies of chromosomal rearrangements in *Drosophila* have shown that proximity to centromeres, telomeres or heterochromatic blocks may suppress gene expression, presumably by altering the structure of a large chromatin domain. Fascioscapulohumeral muscular dystrophy (FSHD; MIM 158900) is a possible example of a similar position effect in man. The gene for this autosomal dominant progressive neuromuscular disease maps close to the telomere of chromosome 4q. When Southern blots of *Eco*RI-digested DNA are hybridized to a subtelomeric probe p13H-11, a very large hybridizing fragment of over 30 kb is seen. DNA from FSHD patients consistently shows smaller bands of 14–28 kb. These patients have a reduced copy number of a 3.2 kb repetitive sequence that is recognized by the probe, and *de novo* deletion of stretches of the repeats have been observed by FSHD patients with no previous family history.

Hopes that the 3.2 kb sequence would contain the *FSHD* gene, however, have been disappointed. Even though it contains part of a homeodomain, there is no evidence that any part of it is transcribed or expressed. Most probably the *FSHD* gene is located proximally to the tandem 3.2 kb repeats, and the deletions move it closer to the 4q telomere, where it is silenced by a position effect.

Other position effects: Evidence of long-range effects controlling gene expression over large chromosomal domains has emerged from studies of disease-associated chromosome breakpoints in humans. Examples are aniridia (AN1; MIM 106210), which is caused by loss-of-function mutations of the *PAX6* gene on 11p13, and campomelic dysplasia (CMPD1; MIM 211970), which is caused by mutations in the *SOX9* gene on 17q24. In each case, affected patients are known who have clearly causative chromosomal breaks, but the breakpoints may be very distant (hundreds of kilobases) from the gene whose expression is affected and do not physically disrupt it. It seems likely that expression of the gene is suppressed by a long-range effect analogous to the classic position effects described above, reflecting the novel chromosomal environment created by the translocation.

Prader-Willi and Angelman syndromes bring together position effects, imprinting and DNA methylation. A *cis*-acting sequence analogous to the globin locus control region has been identified which governs parent-specific methylation and gene expression of a megabase-size chromosomal region at 15q11.

X inactivation: X chromosome inactivation in mammals appears to be initiated by a single gene, *XIST*, which is uniquely expressed on the inactivated X chromosome. This effect is not understood but must be mediated by some sort of long-range chromatin structural change. This is so because a diffusible *XIST*-mediated agent would not be able to affect just the X chromosome on which the *XIST* gene is expressed.

Expression of individual genes in gene clusters may be coordinated by a common locus control region

Some human gene clusters show evidence of coordinated expression of the individual genes in the cluster. For example, individual genes in the α-globin, β-globin and the four *HOX* gene clusters are activated sequentially in a temporal sequence that corresponds exactly with their linear order on the chromosome. In the case of the globin genes, there is a clear developmental stage-specific expression: different genes can be active at the embryonic, fetal or adult stages to generate slightly different forms of hemoglobin.

Recently, it has become apparent that the expression of the genes in each of the two human globin gene clusters is coordinated by a dominant control region, the locus control region (LCR) which is located some distance upstream of the gene cluster. Such cluster-specific LCRs are thought to organize the cluster into an active chromatin domain and to act as enhancers of globin gene transcription. The open conformation of transcriptionally active chromatin domains makes them more accessible to cleavage by the enzyme DNase I. Consistent with this relationship, the β-globin LCR has been considered to comprise short sequences at three major erythroid-specific DNase I-hypersensitive sites (HS2, HS3 and HS4) clustered over a 15 kb region located about 50–60 kb upstream of the β-globin gene, while the α-globin LCR has been identified to occur at an erythroid-specific DNase-hypersensitive site, HS-40, located 60 kb upstream of the α-globin gene. Each site marks the location of what is effectively an *enhancer* sequence of about 200–300 bp of DNA which contains short *cis*-acting sequence elements, including multiple sequence elements recognized by erythroid-specific transcription factors.

Without the respective LCRs, globin gene expression is negligible and, in the case of the β-globin LCR, it appears that the HS2, HS3 and HS4 elements interact with each other to form a larger complex that interacts with the individual globin genes.

Other DNase I-hypersensitive sites are located at the promoters of the globin genes, but show developmental stage specificity. For example, in fetal liver, the promoters of the two γ genes, the β and δ genes, are marked by DNase I-hypersensitive sites but, in adult bone marrow, the two γ genes are no longer transcriptionally active and their promoters no longer reveal DNase I-hypersensitive sites.

Developmental stage-specific switching in globin gene expression is then thought to be accomplished by competition between the globin genes for interaction with their respective LCR and stage-specific activation of gene-specific silencer elements. For example, transcription of the å-globin gene (*HBE1*) is preferentially stimulated by the neighboring LCR at the embryonic stage. In the fetus, however, å-globin expression is suppressed following activation of a silencer and γ-globin expression becomes dominant.

In addition to the human globin LCRs a number of different additional LCRs have been identified (see Kioussis and Festenstein, 1997). However, the role of the human β-globin LCR (which has relied on analysis of human transgenes) has been challenged by gene targeting studies in mice which show that the β-globin LCR has a contributory function rather than a dominant one, and that it is not required for initiation of DNase sensitivity and expression of the genes in the mouse β-globin cluster. These contradictory findings may possibly mean that there is a functional difference between the human and murine LCRs.

Some human genes show selective expression of only one of the two parental alleles

X-linked genes in females and all autosomal genes are biallelic because both father and mother normally contribute one allele each. In males possessing one X chromosome and one Y chromosome, the great majority of sex-linked genes are monoallelic: most of the many genes on the X do not have a functional homolog on the Y chromosome; and some of the few genes on the Y chromosome are known to be Y-specific, for example *SRY*, the major male sex-determining locus. A few genes on the Y chromosome do have functional homologs on the X chromosome and so are biallelic. In some cases of X-Y homologous loci, both homologs are normally functional.

We are accustomed to assuming that both the paternal and maternal alleles of biallelic genes are expressed, unless one or both copies have sustained mutations which affect expression. Clearly the expression can be tissue-specific so that in some cells both parental alleles are strongly expressed; in others, both gene copies are not apparently expressed.

Thus, although there may be cell type-specific differences in expression, there is no discrimination between the capacity of the two parental alleles to be expressed, other than that due to genetic (mutational) differences between them. However, in humans and other mammals, several biallelic genes are known where the expression of one parental allele, either the paternal or the maternal allele but not both, is *normally* repressed in some cells (allelic exclusion).

In such cells the relevant gene is said to exhibit functional hemizygosity: only one half of the maximum gene product is normally obtained *even although the sequences of both parental alleles are perfectly consistent with normal gene expression and may even be identical.* In some cases the allelic exclusion may be a property of select cells or tissues while in other cells of the same individual both alleles may be expressed normally.

Although initially considered a rarity, monoallelic expression of biallelic genes has been demonstrated for a growing number of human genes. A variety of different expression mechanisms can be involved and two broad classes of mechanism are involved (see Chess, 1998; Ohlsson *et al.*, 1998):

- *Allelic exclusion according to parent of origin (imprinting).* In some cases, the choice of which of the two inherited copies is expressed is not random. This means that for some genes the allele whose expression is repressed is always the paternally inherited allele; in others it is always the maternally inherited allele.
- *Allelic exclusion independent of parent of origin.* Here the decision as to which of the two alleles is repressed is initially made randomly, but afterwards that pattern of allelic exclusion is transmitted stably to daughter cells following cell division. A variety of different mechanisms may be involved. In some cases, complex gene regulation may be required. For example, olfactory receptor genes are found in large clusters or arrays in mammalian genes. In an individual olfactory neuron only one allelic array of olfactory receptor genes is active. A unique form of control is the programed DNA rearrangements which are required for individual cell-specific expression of the immunoglobulin genes in B cells and the T cell receptor genes in T lymphocytes.

Genomic imprinting involves differences in the expression of alleles according to parent of origin

Various observations in mammals have suggested that the maternal and paternal genomes in an individual are not equivalent. In addition to genetic differences between the DNA of the sperm and oocyte genomes, there are also epigenetic differences.

A major difference is in both the total amount of DNA methylation (the sperm genome is more extensively methylated than the oocyte genome) and the pattern of DNA methylation in specific DNA sequence classes. For example, Line 1 sequences are highly methylated in sperm cells but only partially methylated in the oocyte (Razin and Kafri, 1994; Yoder *et al.*, 1997). At some individual gene loci, too, there are major differences between the extent of methylation of paternal and maternal alleles. For example, the paternal allele of the ***H19*** gene is heavily methylated; the maternal allele is undermethylated.

Differences between the paternal and maternal genomes lead to differences in expression between paternal and maternal alleles. Genomic imprinting (also called gametic or parental imprinting) in mammals describes the situation where there is nonequivalence in expression of alleles at certain gene loci, dependent on the parent of origin (Reik and Walter, 1998; Brannan and Bartolomei, 1999; Tilghman, 1999). In all (or at least some) of the tissues where the gene is expressed, the expression of either the paternally inherited allele or the maternally inherited allele, is consistently repressed, resulting in monoallelic expression. The same pattern of monoallelic expression can be faithfully transmitted to daughter cells following cell division. However, as the nucleotide sequence of the allele whose expression is repressed may be perfectly consistent with gene expression (and may even be identical to that of the expressed allele), this is an epigenetic phenomenon, not a genetic one.

Prevalence and evolution of imprinting: Most human genes are not subject to imprinting, otherwise we would not see so many simple mendelian characters. Systematic surveys have been made to identify imprinted chromosomal regions in the mouse. Unlike in humans, all mouse chromosomes are acrocentric and Robertsonian translocations can permit crosses to be set up which produce offspring having both copies of one particular chromosome derived from a single parent. These reveal that UPD for some chromosomes has no phenotypic effect; for others it produces abnormal phenotypes. The abnormal phenotypes are sometimes complementary for different parental origins, e.g. overgrowth is often seen in maternal UPD and growth retardation in paternal UPD. For some chromosomes, UPD is lethal.

Further dissection at the chromosomal and genetic levels shows that imprinting is a property of a limited number of individual genes or small chromosomal regions. Currently, a total of over 30 genes are known to be imprinted in humans and mice (electronic reference 1), but the list can be expected to grow. Thus far, two major clusters of imprinted genes are known in the human genome: a 1 Mb region at 11p15 (encompassing the Beckwith- Wiedemann region) contains at least seven imprinted genes which may be arranged in two clusters (Lee *et al.*, 1999); a 2.3 Mb cluster at 15q11-q13 region (encompassing the Prader-Willi and Angelman syndrome loci) also contains at least seven imprinted genes. The imprinted gene clusters contain examples of neighboring genes with different parental imprints e.g. the *H19* gene is expressed only from the maternal chromosome 11 whilst the adjacent *IGF2* gene is expressed only from the paternal chromosome.

The great majority of known imprinted genes are autosomal. However, the *XIST* gene which has a major role in establishing X chromosome inactivation (see next section) may be considered an example of an imprinted X-linked gene since expression of the maternally inherited allele is preferentially repressed in trophoblast. An imprinted X-linked gene which affects cognitive function has also been suggested from differential behavior patterns in Turner syndrome. Girls with Turner syndrome lack a Y chromosome but have only one X chromosome. If the X chromosome is inherited from the mother, socially disruptive behavior is common, but if inherited from the father, the girl shows behavior closer to normal for that of a girl (Skuse *et al.*, 1997).

Imprinting is known to occur in seed plants, some insects and mammals. No major imprinting effect, as judged by phenotype, has been observed in some model organisms such as *Drosophila*, *C. elegans* and the zebrafish, although the potential for imprinting may exist in *Drosophila*. Mammals are unusual in the way in which embryos are totally dependent on flow of nutrients from the maternal placenta. As many imprinted genes are involved in regulating fetal growth, one explanation envisages conflict between the parental genomes: the paternal genome propagates itself best by creating an embryo which aggressively removes nutrients from the mother; the maternal genome suppresses this to protect the mother and spare some resources for future offspring. As seen in cases of uniparental diploidy, paternal genes are preferentially expressed in the trophoblast and extraembryonic membranes, while maternal genes are preferentially expressed in the embryo.

The mechanism of genomic imprinting is unclear but a key component appears to be DNA methylation

To confirm imprinting of a gene, it is necessary to identify an individual who is heterozygous for a sequence variant present in the mature mRNA. mRNA from different tissues can then be checked for monoallelic or biallelic expression, and the origin of each allele determined by typing the parents. For some genes, this type of analysis has shown that imprinting is confined to only certain tissues or to certain stages of development. Thus imprinting allows an extra level of control of gene expression, but it is not possible to compress its functioning into a simple uniform story.

The above observations suggest that some mechanism must be able to distinguish between maternally and paternally inherited alleles: as chromosomes pass through the male and female germlines they must acquire some imprint to signal a difference between paternal and maternal alleles in the developing organism. A key component, at least in maintaining the imprinted status, is allele-specific DNA methylation. The imprinting of several imprinted genes has been shown to be disrupted in mutant mice that are deficient in the *Dnmt1* cytosine methyltransferase gene and all imprinted genes are characterized by CG-rich regions of differential methylation.

Intriguingly, *Dnmt1* is known to have sex-specific exons. In oocytes this results in an oocyte-specific amino-terminal truncated protein product which conceivably could specifically methylate the maternal alleles of genes such as the insulin-like growth factor II receptor. The spermatocyte-specific exon of *Dnmt1* interferes with translation of *Dnmt1* mRNA, and it is less clear how paternal-specific patterns of methylation could be acquired. During development, the imprint would be expected to be stably inherited at least for many rounds of DNA duplication (but see below). Clearly, there must also be a mechanism for erasing the imprints during transmission through the germline, as required when, for example, a man passes on an allele which he has inherited from his mother. Again, however, one can envisage the demethylation that occurs during the early embryo as one way of achieving this, leaving the primordial germ cells essentially unmethylated.

The timing of imprinting has become more clear recently. In the female germline, the maternal imprint, including the maternal pattern of methylation, is likely to be established during oocyte maturation which is consistent with the finding that the Dmnt1 protein is not detectable in nongrowing oocytes but is produced abundantly in growing oocytes. In the male germline, the functional paternal imprint is likely to be established prior to meiosis, possibly in the postmitotic primary spermatocyte.

Imprinted genes frequently reside in clusters with genes expressed on opposite chromosomes often located next to each other, and often containing genes which appear to encode a mature RNA. Adjacent genes appear to be jointly regulated. In the case of the Prader-Willi/Angelman syndrome cluster on 15q11-q13, for example, a single region adjacent to the *SNRPN* gene, termed the imprinting center, is the dominant regulatory sequence and appears to act over comparatively large distances. However, different mechanisms may be found in different imprinted clusters, or even within a single cluster. For example, the mouse *H19, Igf2* and *Ins2* genes are jointly regulated, sharing two endodermal-specific enhancers that are located 3' to *H19*. But other genes in the cluster are not subject to this control, suggesting that multiple control mechanisms may occur within an imprinted gene cluster. Different imprinting mechanisms have been considered in relation to the *H19/Igf2* regulation, including enhancer competition, but to explain a variety of contradictory findings, an imprinting center adjacent to *H19* has been considered to function as a chromatin boundary (insulator) element.

X chromosome inactivation in mammals involves very long-range cis-acting repression of gene expression

Nature of X chromosome inactivation: X chromosome inactivation is a process that occurs in all mammals, resulting in selective inactivation of alleles on one of the two X chromosomes in females (Migeon, 1994; Lyon, 1999). It provides a mechanism of dosage compensation which overcomes sex differences in the expected ratio of autosomal gene dosage to X chromosome gene dosage (which is 2:1 in males but 1:1 in females). Males with a single X chromosome are constitutionally hemizygous for X chromosome genes, but females become functionally hemizygous by inactivating one of the parental X chromosome alleles. Not all genes on the X chromosome are subject to inactivation; genes which escape X-inactivation include ones where there is a functional homolog on the Y chromosome, and some genes where gene dosage does not seem to be important.

In rare individuals with an abnormal number of X chromosomes (45,X; 47,XXX; 47, XXY, etc), a single X chromosome remains active no matter how many are present. By contrast, in triploid individuals either one or two X chromosomes remain active and in tetraploids two X chromosomes remain active. Thus, there must be some kind of *counting mechanism* to ensure that one X chromosome remains active for every two sets of autosomes.

In mammals, both X chromosomes are active in the early female embryo. X-inactivation occurs at an early stage in development, being initiated at the late blastula stage in mice, and most likely also in humans. In each cell that will give rise to the female fetus, one of the two parental X chromosomes is *randomly* inactivated (*note* that trophoblast cells are an exception; the paternal X chromosome is preferentially inactivated, which is a classical example of tissue-restricted imprinting). After the paternal or maternal X chromosome is inactivated in a cell, the same X chromosome usually remains inactive in all progeny cells, that is the X chromosome inactivation pattern is clonally inherited. This means that female mammals are mosaics, comprising mixtures of cell lines in which the paternal X is inactivated and cell lines where the maternally inherited X is inactivated. In addition to X chromosome inactivation in female somatic cells, the X chromosome is known to be inactivated transiently during gametogenesis in both males and females.

Mechanism of X chromosome inactivation: The process of X chromosome inactivation is complex, and distinct molecular mechanisms are involved in initiation of inactivation and maintenance of the inactivation.

The X-inactivation center (Xic), which in humans is located at Xq13, controls the initiation and propagation of X-inactivation. At this centre, the *XIST* gene (called *Xist* in rodents) encodes a mature 15 kb RNA product which is uniquely encoded by the inactive X chromosome. *XIST/Xist* is therefore another example of a gene that is subject to monoallelic expression. In the cells of the early embryo, the decision regarding which X chromosome to inactivate is made randomly, and so the allelic exclusion which *XIST/Xist* shows in these cells is independent of parent of origin.

XIST/Xist is essential for Xic function in initating X chromosome inactivation but is not required for maintaining X chromosome inactivation. Somehow *cis*-limited spreading of this RNA product acts so as to coat the inactivated X chromosome over very long distances.

In rodents, coating of the *Xist* RNA gives the inactivated X chromosome a banded pattern suggesting a preferential association with gene-rich Giemsa-minus regions. However, the mechanism of ensuring inactivation of genes on the inactive X but not on the active X is unknown (see Duthie *et al.*, 1999 for possible models).

Although the *Xist* gene is essential for Xic function, *Xist* alone is not sufficient. The X-controlling element (Xce) affects the choice of which X chromosome remains active, and is distinct from *Xist,* being located 3' to it. In addition, deletion of a 65-kb region 3' to *Xist* produces an effect which suggests that elements involved in the counting mechanism lie distal and 3' to *Xist.*

Recently, another gene has been identified as being transcribed from the opposite strand to that which is used for transcribing the *Xist* gene. Because the transcription unit of the new gene completely overlaps the *Xist* gene, and is in the reverse orientation, it has been named *Tsix* (Lee *et al.*, 1999).

This has given rise to the idea that *Xist* may be regulated by the *Tsix* gene (see Heard *et al.*, 1999 for possible models of *Tsix* regulation).

The unique organization and expression of Ig and TCR genes

The organization and expression of immunoglobulin (Ig) and T-cell recepter (TCR) genes is in many ways quite different from that of other genes. This is so because of the need for each individual to produce a huge variety of different Igs and TCRs. An individual B or T lymphocyte is *monospecific* and produces a single type of Ig or TCR; it is the population of different B and T cells in any one individual that enables the synthesis of so many different types of these molecules.

B and T lymphocytes need to be extremely diverse because they represent the cells that provide antibody responses or cell-mediated responses to foreign antigen: by providing a large repertoire of Igs and TCRs, the possibilities for being able to recognize and bind very many different types of foreign antigen are greatly increased.

Ig and TCR genes exhibit a unique organization: multiple gene segments can encode each of several different regions of the polypeptide

Polypeptide structure: An Ig molecule consists of four polypeptide chains, two identical heavy chains and two identical light chains.

The light chains fall into two classes: kappa (ê) and lambda (λ) light chains, which are functionally equivalent. At the N-terminal segments of each type of chain are the so-called variable (V) regions, which need to bind foreign antigen; the remaining C-terminal segments are constant (C) regions.

In the case of the heavy chains, there are different alternatives for the constant region which specify the tissues in which the Ig will be expressed and dictate the immunoglobulin class. Similarly, TCRs, which provide cell-mediated immune responses to foreign antigens, consist of two types of chain. Each such chain has Ig-like variable regions which bind foreign antigen, and constant regions which anchor the molecule to the cell surface. The most frequently occurring TCRs have a β and a γ chain; a minor population consists of an α chain and a δ chain.

Gene structure: The genes which encode the different types of chain in Igs and TCRs are located on different chromosomes and are organized as clusters of numerous gene segments. Each such cluster is unusual in that the coding sequences for *specific segments* of each chain are often present in numerous different copies that are sequentially repeated.

For example, although the constant region of human ê light chain Ig is encoded by a single $C_{ê}$ sequence, the variable regions are encoded by a combination of a $V_{ê}$ segment (which encodes most of the variable region) and a short $J_{ê}$ segment (joining segment; encodes a small part at the C-terminal end of the variable region) which are selected from a total of about 76 alternative $V_{ê}$ segments and five alternative $J_{ê}$ segments. Although the λ light chain is similarly encoded by V_{λ}, J_{γ} and C_{γ} segments, the heavy chain Ig locus shows some differences. The variable region is encoded by a combination of a V_H gene segment, a J_H gene segment and also a D_H gene segment (encoding a diversity segment), each selected from many repeated gene segments.

Additionally, there are a variety of different C_H sequences which specify the *class* of the Ig. In total this cluster comprises about 140 gene segments, of which about one-third are known to be incapable of expression, and spans about 1200 kb.

As each Ig gene cluster or TCR gene cluster in an individual B or T lymphocyte only ever gives rise to at most one Ig or TCR polypeptide, an entire cluster can functionally be regarded as a single, albeit unusual, type of gene. However, the individual gene segments cannot be regarded as the functional equivalent of *classical exons.*

This is so because individual gene segments in these clusters are sometimes composed of coding DNA and noncoding DNA and may consist of several exons. For example, each of the human C_H sequences is itself composed of three or four classical exons separated by introns: after transcription into RNA, the intronic sequences are discarded, and only the exonic sequences are retained in the mRNA.

Programmed DNA rearrangements at the Ig and TCR loci occur during the maturation of B and T lymphocytes, respectively

The unique arrangement of gene segments in the Ig and TCR gene clusters reflects the very unusual way in which somatic recombinations are required in B and T lymphocytes before functional Ig and TCR genes can be assembled and then expressed (see below).

Such somatic recombinations result in bringing together different combinations of the different gene segments in different individual lymphocytes. Consequently, they can be regarded as both *tissue-specific* (confined to B and T lymphocytes) and *cell-specific* events which involve *alternative DNA splicing* (as opposed to alternative RNA splicing which brings about different combinations of exons *at the RNA level*).

As a result, the original germline gene organization is altered: gene segments that were distant in the germline are spliced together at the DNA level.

Because the choice of which of the many repeated gene segments are recombined to give a functional V-J or V-D-J unit is *cell specific*, individual B and T cells produce different Igs and TCRs. This means that, in a sense, every individual is a mosaic with respect to the organization of the Ig and TCR genes in B and T lymphocytes; even identical twins will diverge genetically.

The rearrangements which lead to the production of functional light chains and heavy chains of Igs are slightly different.

- *Making a light chain*. In order to generate a functional ê light chain Ig, for example, a somatic recombination event brings together a specific combination of one of the $V_ê$ gene segments and one of the $J_ê$ gene segments (V-J joining). Thereafter, splicing to the single $C_ê$ sequence occurs *at the RNA level*.

- *Making a heavy chain.* Two successive somatic recombinations are required, resulting first in D_H-J_H joining, and then V_H-D_H-J_H joining. Subsequently the resulting V_H-D_H-J_H coding sequence is spliced *at the RNA level* to the nearest C_H sequence, initially C_i.

Because there are three types of functional Ig gene loci in human cells (heavy chain, ê light chain and λ light chain), and because these occur on both maternal and paternal homologs, there are six chromosomal segments in which DNA rearrangments can result in production of an Ig chain.

However, an individual B cell is *monospecific*: it produces only one type of Ig molecule with a single type of heavy chain and a single type of light chain. This is so for two reasons:

- *Allelic exclusion.* A light chain or a heavy chain can be synthesized from a maternal chromosome or a paternal chromosome in any one B cell, but not from both parental homologs. As a result, there is monoallelic expression at the heavy chain gene locus in B cells. This phenomenon also applies to TCR gene clusters.

- *Light chain exclusion.* A light chain synthesized in a single B cell may be a ê chain, or a λ chain, but never both. As a result of this requirement, plus that of allelic exclusion, there is monoallelic expression at one of the two functional light chain gene clusters and *no expression* at the other. The decision to choose which of the two heavy chain alleles and which of the four possible segments to make a light chain appears to be random. Most likely, in each B-cell precursor, productive DNA rearrangements are attempted at all six Ig alleles but the chances of productive arrangements in more than one light chain cluster or more than one heavy chain allele may not be high. Additionally, however, there appears to be some kind of negative feedback regulation: a functional rearrangement at one of the heavy chain alleles suppresses rearrangements occurring in the other allele, and a functional rearrangement at any one of the four regions capable of encoding a light chain suppresses rearrangements occurring in the other three.

V-J and V-D-J joining is often achieved by intrachromatid deletions, and also by megabase inversions in the former case

The genetic mechanism leading to V-J and V-D-J joining often involves large-scale deletions which are thought to occur by an intrachromatid recombination event, similar to those used in V-D-J-C joining (see next section). In addition, V-J joining often occurs as a result of megabase inversions. The human ê light chain gene locus spans about 1840 kb on 2p12 and includes about 76 $V_ê$ segments, mostly comprising pairs of duplicated V gene segments, organized as two clusters: a *proximal cluster* located adjacent to the $J_ê$ segments and to the $C_ê$ segment, and a *distal cluster*.

This occurs as a result of an inverted repeat structure: V gene segments in the proximal $V_ê$ cluster usually have a corresponding duplicate in a distal $V_ê$ cluster which is separated from the proximal cluster by about 800 kb and in the opposite orientation. Depending on which V segment cluster is involved, V-J joining occurs by two possible routes:

- V segments in the distal cluster are joined to J segments by inversions.
- V segments in the proximal cluster are joined to J segments by deletions.

Note that the joining process is imprecise, and so can also introduce a measure of variability in the sequence at the junctions of joined segments.

Class switching of heavy chains involves differential joining of a single VDJ unit to alternative DNA segments encoding constant regions

Although a B cell produces only one type of Ig molecule, the heavy chain class can change during the cell lineage (class switching or isotype switching).

Such switching involves differential joining of the same VDJ unit that was brought together by two successive somatic recombinations to different segments encoding alternative constant regions. The initial joining of a VDJ sequence to constant region segments is accomplished *at the RNA level.* However, subsequently, class switching involves joining the same VDJ unit *at the DNA level* to alternative constant regions by yet more somatic recombination events (V-D-J-C joining). Class switching involves the following progression:

initial synthesis of IgM only by immature B cells. This occurs because the VDJ unit is spliced at the RNA level to a $C_ì$ sequence.

Later synthesis of both IgM and IgD by immature B cells. The partial switch to making IgD occurs because the VDJ unit can be spliced at the RNA level to a C_δ sequence, as a result of alternative RNA splicing.

Synthesis of IgG, IgE or IgA by mature B cells. Class switching events involve splicing the same VDJ unit to a C_γ, $C_å$ or C_α sequence, respectively, at the DNA level as a result of a somatic recombination event (VDJ-C joining). The mechanism involves deletion of the intervening sequence by intrachromatid recombination.

6

INSTABILITY OF THE HUMAN GENOME: MUTATION AND DNA REPAIR

AN OVERVIEW

As in other genomes, the DNA of the human genome is not a static entity. Instead, it is subject to a variety of different types of heritable change (mutation). Large-scale chromosome abnormalities involve loss or gain of chromosomes or breakage and rejoining of chromatids. Smaller scale mutations can be grouped into different mutation classes and can also be categorized on the basis of whether they involve a single DNA sequence or whether they involve *exchanges* between two allelic or nonallelic sequences. Three classes of small-scale mutation can be distinguished:

- Base substitutions - involve replacement of usually a single base; in rare cases several clustered bases may be replaced simultaneously as a result of a form of gene conversion.
- Deletions - one or more nucleotides are eliminated from a sequence.
- Insertions - one or more nucleotides are inserted into a sequence. In rare cases this involves transposition from another locus. Copy or duplicative transposition involves a sequence from one locus being replicated and the copy inserted into another locus. Noncopy transposition involves simple transposition of a DNA

sequence from one locus to another. In human and mammalian genomes, noncopy transposition is very rare: the great majority of DNA transposition occurs via an RNA intermediate so that the insertion is of a sequence copied from another locus.

New mutations arise in single individuals, in somatic cells or in the germline. If a germline mutation does not seriously impair an individual's ability to have offspring who can transmit the mutation, it can spread to other members of a (sexual) population. Allelic sequence variation is traditionally described as a DNA polymorphism if more than one variant (allele) at a locus occurs in a human population with a frequency greater than 0.01 (a frequency high enough such that an origin as a result of chance recurrence is highly unlikely). The mean heterozygosity for human genomic DNA is thought to be of the order of 0.001–0.004 (i.e. approximately 1:250 to 1:1000 bases are different between allelic sequences; Cooper *et al.*, 1985; Nickerson *et al.*, 1998; Taillon-Miller *et al.*, 1998). Certain genes, notably some HLA genes, are exceptionally polymorphic and alleles can show very substantial sequence divergence. Because mutation rates are comparatively low the vast majority of the differences between allelic sequences within an individual are inherited, rather than resulting from *de novo* mutations.

Mutations are the raw fuel that drives evolution, but they can also be pathogenic. They can be the direct cause of a phenotypic abnormality or they can result in increased susceptibility to disease.

The usually low level of mutation may therefore be viewed as a balance between permitting occasional evolutionary novelty at the expense of causing disease or death in a proportion of the members of a species. Normally, most mutations arise as copying errors during DNA replication because DNA polymerases, like all enzymes, are error-prone. The error rate of a DNA polymerase (that is, the frequency of incorporating a wrong base) is significantly reduced by having a subunit of the polymerase which has a proofreading function. Even then, however, the size of the human genome makes huge demands on the fidelity of any DNA polymerase: a sequence of 3 billion nucleotides needs to be replicated accurately every single time a human cell divides.

DNA is also subject to significant spontaneous chemical attack in the cell. For example, every day approximately 5000 adenines or guanines are lost from the DNA of each nucleated human cell by depurination (the N-glycosidic bond linking the purine residue to the carbon 12 of the deoxyribose is hydrolyzed and the purine is replaced by a hydroxyl group at carbon 12). DNA is also damaged by exposure to natural ionizing radiation and to reactive metabolites. In order to minimize the mutation rate, therefore, it is necessary to have effective DNA repair systems which identify and correct many abnormalities in the DNA sequence. In addition, errors that arise in the mRNA sequence during gene expression are subject to RNA surveillance mechanisms which ensure removal of mRNAs which have inappropriate termination codons.

Simple mutations

Mutations due to errors in DNA replication and repair are frequent

Mutations can be induced in our DNA by exposure to a variety of mutagens occurring in our external environment or to mutagens generated in the intracellular environment. In the case of radiation-induced mutation, for example, Dubrova *et al.* 1996 reported that the normal germline mutation rate for hypervariable minisatellite loci was doubled as a consequence of heavy exposure to the radiaoctive fallout from the Chernobyl accident. However, under normal circumstances by far the greatest source of mutations is from *endogenous* mutation, notably spontaneous errors in DNA replication and repair. During an average human lifetime there are an estimated 10^{17} cell divisions: about 2×10^{14} divisions are required to generate the approximately 10^{14} cells in the adult, and additional mitoses are required to permit cell renewal in the case of certain cell types, notably epithelial cells (see Cairns, 1975). As each cell division requires the incorporation of 6×10^{9} new nucleotides, error-free DNA replication in an average lifetime would require a DNA replication-repair process with an accuracy great enough so that the correct nucleotide was inserted on the growing DNA strands on each of about 6×10^{26} occasions.

Such a level of DNA replication fidelity is impossible to sustain; indeed, the observed fidelity of replication of DNA polymerases is very much less than this and uncorrected replication errors occur with a frequency of about 10^{-9}-10^{-11} per incorporated nucleotide. As the coding DNA of an average human gene is about 1.7 kb, coding DNA mutations will occur spontaneously with an average frequency of about 1.7×10^{-6}-1.7×10^{-8} per gene per cell division. Thus, during the approximately 10^{16} mitoses undergone in an average human lifetime, each gene will be a locus for about 10^{8}-10^{10} mutations (but for any one gene, only a tiny minority of cells will carry a mutation). In many cases, a deleterious gene mutation in a somatic cell will be inconsequential: the mutation may cause lethality for that single cell, but will not have consequences for other cells. However, in some cases, the mutation may lead to an inappropriate continuation of cell division, causing cancer.

The frequency of individual base substitutions is nonrandom

Base substitutions are among the most common mutations and can be grouped into two classes:

- Transitions are substitutions of a pyrimidine (C or T) by a pyrimidine, or of a purine (A or G) by a purine.
- Transversions are substitutions of a pyrimidine by a purine or of a purine by a pyrimidine.

When one base is substituted by another, there are always two possible choices for transversion, but only one choice for a transition. For example, the base adenine can undergo two possible transversions. One might, therefore, expect transversions to be twice as frequent as transitions. Because the substitution of alleles in a population takes thousands or even millions of years to complete, nucleotide substitutions cannot be observed directly. Instead, they are always inferred from pairwise comparisons of DNA molecules that share a common origin, such as orthologs in different species. When this is done, the transition rate in mammalian genomes is found to be unexpectedly higher than transversion rates. For example, Collins and Jukes (1994) compared 337 pairs of human and rodent orthologs and found that the transition rate consistently exceeded the transversion rate. The ratio was 1.4 to 1 for substitutions which did not lead to an altered amino acid, and more than 2 to 1 for those that did result in an amino acid change.

Transitions may be favored over transversions in coding DNA because they usually result in a more conserved polypeptide sequence. In both coding and noncoding DNA the excess of transitions over transversions is at least partly due to the comparatively high frequency of C T transitions, resulting from instability of cytosine residues occurring in the CpG dinucleotide. In such dinucleotides the cytosine is often methylated at the 5' C atom and 5-methylcytosines are susceptible to spontaneous deamination to give thymine. Presumably as a result of this, the CpG dinucleotide is a hotspot for mutation in vertebrate genomes: its mutation rate is about 8.5 times higher than that of the average dinucleotide. Other factors favoring transitions over transversions are likely to include differential repair of mispaired bases by the sequence-dependent proofreading activities of the relevant DNA polymerases.

The frequency and spectrum of mutations in coding DNA differs from that in noncoding DNA

Many mutations are generated essentially randomly in the DNA of individuals. As a result, coding DNA and noncoding DNA are about equally susceptible to mutation. Clearly, however, the major consequences of mutation are largely restricted to the approximately 3% of the DNA in the human genome which is coding DNA. Mutations which occur in this component of the genome are of two types:

- Silent (synonymous) mutations do not change the sequence of the gene product.
- Nonsynonymous mutations result in an altered sequence in a polypeptide or functional RNA: one or more components of the sequence are altered or eliminated, or an additional sequence is inserted into the product.

Silent mutations are thought to be effectively *neutral* mutations (conferring no advantage or disadvantage to the organism in whose genome they arise). In contrast, nonsynonymous mutations can be grouped into three classes, depending on their effect: those having a deleterious effect; those with no effect; and those with a beneficial effect (e.g. improved gene function or gene-gene interaction). Most new nonsynonymous mutations are likely to have a deleterious effect on gene expression and so can result in disease or lethality. However, the frequency of such mutation in the population is very much reduced because of natural selection. As a result, the overall mutation rate in coding DNA is much less than that in noncoding DNA. Consequently, the coding DNA component of a specific gene and the derived amino acid sequence show a relatively high degree of evolutionary conservation, as do important regulatory sequences such as the multiple elements of promoters and enhancers, and intronic sequences immediately flanking exons.

Selection pressure (the constraints imposed by natural selection) reduces both the overall frequency of surviving mutations in coding DNA and the spectrum of mutations seen. For example, deletions/insertions of one or several nucleotides are frequent in noncoding DNA but are conspicuously absent from coding DNA. This is so because often such mutations will cause a shift in the translational reading frame (frameshift mutation), introducing a premature termination codon and causing loss of gene expression. Even if insertions/deletions do not cause a frameshift mutation, they can often affect gene function, for example, as a result of removing a key coding sequence. Instead, coding DNA is marked by a comparatively high frequency of nonrandom base substitution occurring at locations which lead to minimal effects on gene expression.

The location of base substitutions in coding DNA is nonrandom

Nucleotide substitutions occurring in noncoding DNA usually have no net effect on gene expression. Exceptions include some changes in promoter elements or some other DNA sequence that regulates gene expression, and in important intronic sequence positions, such as at splice junctions or the splice branch site. Substitutions occurring in coding DNA sequences which specify polypeptides show a very nonrandom pattern of substitutions because of the need to conserve polypeptide sequence and biological function. In principle, base substitutions can be grouped into three classes, depending on their effect on coding potential.

The different classes of base substitution listed in the box show differential tendencies to be located at the first, second or third base positions of codons. Because of the design of the genetic code, different degrees of degeneracy characterize different sites. Base positions in codons can be grouped into three classes:

- Nondegenerate sites are base positions where all three possible substitutions are nonsynonymous. They include the first base position of all but eight codons, the second base position of all codons and the third base position of two codons, AUG and UGG. Taking into account the observed codon frequencies in human genes, they comprise about 65% of the base positions in human codons. The base substitution rate at nondegenerate sites is very low, consistent with a strong conservative selection pressure to avoid amino acid changes.
- Fourfold degenerate sites are base positions in which all three possible substitutions are synonymous and are found at the third base position of several codons. They comprise about 16% of the base positions in human codons. The substitution rate at fourfold sites is very similar to that within introns and pseudogenes, consistent with the assumption that synonymous substitutions are selectively neutral.
- Twofold degenerate sites are base positions in which one of the three possible substitutions is synonymous. They are often found at the third base positions of codons, but also at the first base position in eight codons. They comprise about 19% of the base positions in human codons. As expected, the substitution rate for twofold degenerate sites is intermediate: only one out of the

three possible substitutions, a transition, maintains the same amino acid. The other two possible substitutions are transversions which, because of the way in which the genetic code has evolved, are often conservative substitutions. For example, at the third base position of the glutamate codon GAA, a transition A G is silent, while the two transversions (A C; A T) result in replacement by a closely similar amino acid, aspartate.

The design of the genetic code and the degree to which one amino acid is functionally similar to another affect the relative mutabilities of individual amino acids. Certain amino acids may play key roles which cannot be substituted easily by others. For example, cysteine is often involved in disulfide bonding which can play a crucially important role in establishing the conformation of a polypeptide. As no other amino acid has a side chain with a sulfhydryl group, there is strong selection pressure to conserve cysteine residues at many locations, and cysteine is among the least mutable of the amino acids (Collins and Jukes, 1994). In contrast, certain other amino acids such as serine and threonine have very similar side chains, and substitutions at both the first base position of codons (ACX UCX; where X = any nucleotide) and second base positions (ACPy AGPy; where Py = pyrimidine) can result in serine threonine substitutions. Presumably as a result, serine and threonine are among the most mutable of the amino acids.

Protein-coding genes show enormous variation in the rate of nonsynonymous substitutions

The rate and type of substitution varies between different genes. At one extreme are proteins whose sequences are extremely highly conserved, such as ubiquitin, histones H3 and H4, calmodulin, ribosomal proteins, etc. For example, the ubiquitin proteins of humans, mouse and *Drosophila* show 100% sequence identity, and comparison with the yeast ubiquitin reveals 96.1% sequence identity. These genes are not especially protected from mutation, because the rate of synonymous codon substitution is typical of that for many protein-encoding genes. Instead, what distinguishes them is the extremely low rate of nonsynonymous codon substitution compared with other genes.

Presumably, ubiquitin and the other highly conserved proteins play such crucial roles that they are under huge selection pressure to conserve the sequence. At the other extreme, the fibrinopeptides are proteins which are evolving extremely rapidly and do not appear to be subject to any selective constraint.

These proteins (only 20 amino acids long) are thought to be functionless - they are fragments which are generated as part of the protein *fibrinogen* and discarded when the protein is activated to form *fibrin* during blood clotting. Another extremely rapidly evolving sequence is the major sex-determining locus, *SRY*. This gene encodes a protein which contains a central 'high mobility group' domain (HMG box) of about 78 amino acids. The HMG box is central to *SRY* function and is well conserved, but the flanking N- and C-terminal segments are evolving extremely rapidly, which may indicate that the majority of the *SRY* coding sequence is not functionally significant (Whitfield *et al.*, 1993). In between the two extremes in the rate of nonsynonymous substitution are the vast majority of polypeptide-encoding genes.

The molecular clock can vary from gene to gene, and is different in different lineages

Synonymous substitutions have been considered to be effectively neutral from the point of view of selective constraints. As a result, the concept of a constant molecular clock (whereby a given gene or gene product undergoes a constant rate of molecular evolution) was suggested over 30 years ago. Since then, however, abundant evidence has been accumulated which does not support the concept.

When substitution rates are compared for different genes, even closely related members of a gene family, there are considerable differences.

Timing of DNA replication. The DNA of different genomic components is replicated at different times. Actively transcribing genes are replicated early; transcriptionally inactive DNA such as the inactivated X chromosome is replicated late. Early replicating and late replicating DNA may be subject to different intracellular concentrations of free nucleotides and of the various enzymes involved in replication and DNA repair, causing possible differences in mutation rates.

Differences in GC content. This parameter is not independent of the previous one because the early replicating DNA is relatively GC-rich. The present evidence suggests that any relationship between the GC content of a mammalian gene and its mutation rate is not a simple one (Sharp and Matassi, 1994).

Different genomes. The mitochondrial DNA in mammals and many other animals is thought to be evolving at a much higher rate than nuclear DNA.

For a given gene, the molecular clock varies very considerably depending on the species lineage, and the clock runs at different rates for closely related members of a gene family (e.g. Gibbs *et al.*, 1998). In order to estimate the relative rates of nucleotide substitutions in two lineages leading to present-day species A and B, a relative rate test is used. This involves using a third reference species C which is known to have branched off earlier in evolution, before the A-B split. Pairwise comparisons of orthologs in A and C, and in B and C are then used to calculate the K value, the number of synonymous substitutions per 100 sites. The K_{AC} and K_{BC} values then provide a measure of the relative rates of mutation in the lineages leading to species A and to species B. For example, when a variety of orthologs in mouse (species A) and rat (species B) are referenced against orthologs in humans (species C), the overall K_{AC} and K_{BC} values are nearly identical (Li and Graur, 1991, p. 82). This suggests that the base substitution rates in the lineages leading to present-day mouse and rat have been nearly equal. However, similar analyses suggest that the substitution rate appears to be lower in lineages leading to the primates and lower still in the lineage leading to modern day humans.

Higher mutation rates in males are likely to be related to the greater number of germ cell divisions

Since Haldane first observed that most mutations resulting in hemophilia were generated in the male germline, it has been assumed that, at least in humans, mutations are preferentially paternally inherited. Two major approaches have been taken to estimate the relative mutation rates in the male and female germlines:

Molecular evolutionary methods. The starting point for such methods typically investigates known homologous genes on the X and Y chromosomes located outside the pseudoautosomal region. The sequences are compared with orthologs in another species to estimate the rate of synonymous mutations for the X chromosome gene (K_{SX}) and the Y chromosome gene (K_{SY}). Unlike autosomes, the sex chromosomes spend different amounts of time in the two sexes. The great majority of Y chromosome sequences spend all their time in males. By contrast, as females have two X chromosomes whereas males have one X chromosome, X chromosome sequences spend on average 2/3 of their time in females and 1/3 of their time in males. If we use the symbol α to represent the ratio of the male mutation rate to the female mutation rate this is equivalent to setting the male mutation rate at α relative to a female mutation rate of 1. For most Y chromosome sequences, the mutation rate is therefore α. For X chromosome sequences it is 2/3 times the mutation rate in female cells (1) plus 1/3 times the mutation rate in males (α), giving a total of $2/3 + \alpha/3$. Therefore the observed K_{SX}/K_{SY} ratio $= (2/3 + \alpha/3)/\alpha$ and so can be used to estimate α. This results in estimates of 3–10 for α.

Direct observation of disease-causing mutations. Clearly, the mutations assessed in this case are a special subset. The approach involves analysing samples from a patient with a *de novo* mutation and the two parents (the parent who passed on the faulty chromosome is typically identified by typing for markers closely flanking the disease gene). In most cases one would expect the mutation to have been transmitted through the germline (although if only a blood sample is typed, it is possible that some mutations could be postzygotic). In the case of point mutation analyses for disorders where imprinting is not suspected, estimates for α again support a higher mutation rate in males.

The comparatively high male mutation rate may be due to different factors, but a major contributory factor is thought to be the large sex difference in the number of human germ cell divisions. In females, the number of cell divisions from zygote to fertilized oocyte is constant because all of the oocytes have been formed by the fifth month of development and only two further cell divisions are required to produce the zygote.

The estimated number of successive female cell divisions from zygote to mature egg has been variously estimated as 24 (Vogel and Motulsky, 1996) and 31 (Li, 1997, p. 229) and is broadly similar to estimates of 30–31 male cell divisions required from zygote to stem spermatogonia at puberty. Five subsequent cell divisions are required for spermatogenesis but thereafter the spermatogenesis cycle occurs approximately every 16 days or 23 cycles per year.

This means that in males, the number of cell divisions required to produce sperm is age-dependent. If an average age of 13 is taken for onset of puberty and an average of 25 for male reproductive age, the total number of cell divisions is about 30 + 5 + [23 × (25 - 13)], or about 310 divisions. Given that errors in DNA replication/repair provide the great majority of mutations, one might then expect that the male mutation rate would be substantially greater than that of the female.

Genetic mechanisms which result in sequence exchanges between repeats

In addition to very frequent simple mutations, there are several mutation classes which involve sequence exchange between allelic or nonallelic sequences, often involving repeated sequences. For example, tandemly repetitive DNA is prone to deletion/insertion polymorphism whereby different alleles vary in the number of integral copies of the tandem repeat. Such variable number of tandem repeat (VNTR) polymorphisms can occur in the case of repeated units that are very short (microsatellites); intermediate (minisatellites) or large. Different genetic mechanisms can account for VNTR polymorphism depending on the size of the repeating unit (see the following two sections). In addition, interspersed repeats can also predispose to deletions/duplications by a variety of different genetic mechanisms.

Slipped strand mispairing can cause VNTR polymorphism at short tandem repeats (microsatellites)

There is considerable variation in the germline mutation rates at microsatellite loci, ranging from an undetectable level up to about 8×10^{-3} (Mahtani and Willard, 1993; Weber and Wong, 1993). Novel length alleles at (CA)/ (TG) microsatellites and at tetranucleotide marker loci are known to be formed without exchange of flanking markers. This means that they are not generated by unequal crossover (see below). Instead, as new mutant alleles have been observed to differ by a single repeat unit from the originating parental allele (Mahtani and Willard, 1993), the most likely mechanism to explain length variation is a form of exchange of sequence information which commences by slipped strand mispairing. This occurs when the normal pairing between the two complementary strands of a double helix is altered by staggering of the repeats on the two strands, leading to incorrect pairing of repeats. Although slipped strand mispairing can be envisaged to occur in nonreplicating DNA, replicating DNA may offer more opportunity for slippage and hence the mechanism is often also called replication slippage or polymerase slippage.

In addition to mispairing between tandem repeats, slippage replication has been envisaged to generate large deletions and duplications by mispairing between noncontiguous repeats and has been suggested to be a major mechanism for DNA sequence and genome evolution. The pathogenic potential of short tandem repeats is considerable.

Large units of tandemly repeated DNA are prone to insertion/deletion as a result of unequal crossover or unequal sister chromatid exchanges

Homologous recombination describes recombination (crossover) occurring at meiosis or, rarely, mitosis between identical or very similar DNA sequences. It usually involves breakage of *nonsister chromatids* of a pair of homologs and rejoining of the fragments to generate new recombinant strands.

Sister chromatid exchange is an analogous type of sequence exchange involving breakage of individual *sister chromatids* and rejoining fragments that initially were on different chromatids of the same chromosome. Both homologous recombination and sister chromatid exchange normally involve *equal* exchanges - cleavage and rejoining of the chromatids occurs at the same position on each chromatid. As a result, the exchanges occur *between allelic sequences* and at corresponding positions within alleles. In the case of intragenic equal crossover between two alleles, a new allele can result which is a fusion gene (or hybrid gene), comprising a terminal fragment from one allele and the remaining sequence of the second allele. However, equal sister chromatid exchanges cannot normally produce genetic variation because sister chromatids have identical DNA sequences.

Unequal crossover is a form of recombination in which the crossover takes place *between nonallelic sequences* on nonsister chromatids of a pair of homologs. Often the sequences at which crossover takes place show very considerable sequence homology which presumably stabilizes mispairing of the chromosomes. Because crossover occurs between mispaired nonsister chromatids, the exchange results in a deletion on one of the participating chromatids and an insertion on the other. The analogous exchange between sister chromatids is called unequal sister chromatid exchange. Both mechanisms occur predominantly at locations where the tandemly repeated units are moderate to large in size. In such cases, the very high degree of sequence homology between the different repeats can facilitate pairing of nonallelic repeats on nonsister chromatids or sister chromatids. If chromosome breakage and rejoining occurs while the chromatids are mispaired in this way, an insertion or deletion of an integral number of repeat units will result. Note that such exchanges are *reciprocal*; both participating chromatids are modified, in one case resulting in an insertion, and in the other case in a complementary deletion.

Unequal sister chromatid exchange is thought to be a major mechanism underlying VNTR polymorphism in the rDNA clusters. Unequal crossover is also expected to occur comparatively frequently in complex satellite DNA repeats and at tandemly repeated gene loci. In the latter case, unequal crossover is known to generate pathogenic deletions at some loci. Such exchanges can also lead to concerted evolution by causing a particular variant to spread through an array of tandem repeats, resulting in *homogenization* of the repeat units.

Occasionally, unequal crossover and unequal sister chromatid exchanges can occur at regions where there is little homology. This is likely to be the case when such mechanisms first generate a tandemly duplicated locus following mispairing of nonallelic repeats such as two *Alu* repeats or even smaller elements.

Gene conversion events may be relatively frequent in tandemly repetitive DNA

Gene conversion describes a *nonreciprocal* transfer of sequence information between a pair of nonallelic DNA sequences (*interlocus gene conversion*) or allelic sequences (*interallelic gene conversion*). One of the pair of interacting sequences, the *donor*, remains unchanged. The other DNA sequence, the *acceptor*, is changed by having some or all of its sequence replaced by a sequence copied from the donor sequence. The sequence exchange is therefore a directional one; the acceptor sequence is modified by the donor sequence, but not the other way round.

One possible mechanism for gene conversion envisages formation of a heteroduplex between a DNA strand from the donor gene and a complementary strand from the acceptor gene. Following heteroduplex formation, conversion of an acceptor gene segment may occur by mismatch repair - DNA repair enzymes recognize that the two strands of the heteroduplex are not perfectly matched and 'correct' the DNA sequence of the acceptor strand to make it perfectly complementary in the converted region to the sequence of the donor gene strand.

Gene conversion has been well-described in fungi where all four products of meiosis can be recovered and studied (tetrad analysis). In humans and mammals it is not possible to do this and so gene conversion cannot be demonstrated unambiguously in higher organisms (it can never be distinguished from double crossover events, for example, although double crossovers occurring in very close proximity would normally be expected to be extremely unlikely). Despite the difficulty in identifying gene conversion in complex organisms, there are numerous instances in mammalian genomes where an allele at one locus shows a pattern of mutations which strongly resembles those found in alleles at another locus of the same species. Such evidence suggests gene conversion-like exchanges between loci.

Although simple comparisons of two sequences may be suggestive, the evidence for gene conversion is most compelling when a new mutant allele can be compared directly with its progenitor sequence. Certain highly mutable loci lend themselves to this type of analysis. In particular, some hypervariable minisatellite loci have high germline mutation rates (often 1% or more per gamete) and individual repeats often show nucleotide differences so that repeat subclasses can be recognized. Germline mutations can be studied by detecting and characterizing mutant mini-satellite alleles in individual gametes. To do this, PCR analysis has been conducted on multiple dilute aliquots of DNA isolated from the sperm of an individual (small pool PCR), where each aliquot is calibrated to contain a few, perhaps 100, input molecules (Jeffreys *et al.*, 1994).

The PCR products recovered from individual pools can then be typed to identify any new mutations that result in a novel allele whose length is sufficiently different as to be distinguishable from the progenitor allele. Analyses of the patterns of germline mutation at three such loci have failed to identify exchanges of flanking markers and have shown that most mutations occurring at these loci are polar, involving the preferential gain of a few repeats at one end of a tandem repeat array. There is a bias towards gain of repeats and evidence was obtained for nonreciprocal sequence exchange between alleles, suggesting interallelic gene conversion. Evidence for interlocus gene conversion has also been obtained in human genes, notably the steroid 21-hydroxylase gene.

Pathogenic mutations

There is a high deleterious mutation rate in hominids

Neutral mutations (those which are neither detrimental nor advantageous for the organism carrying them) accumulate throughout the generations at a rate equal to the mutation rate. To get an estimate of the (total) mutation rate is therefore simple: all one needs to do is to measure the rate of change of some presumed neutral sequence (e.g. intronic, pseudogene, etc). The deleterious mutation rate, by contrast, has been notoriously difficult to measure and no convincing estimate existed for any vertebrate until a study reported by Eyre-Walker and Keightley (1999). They investigated amino acid changes in 46 proteins occurring in the human ancestral line after its divergence from the chimpanzee. If all non-synonymous substitutions were neutral, 231 new substitutions would have been expected in their sample of genes (given an average neutral mutation rate of 0.0056 nonsynonymous substitutions per nucleotide and a total of 41 471 nucleotides investigated). Instead, only 143 nonsynonymous substitutions were observed and 88 such substitutions were inferred to have been removed by natural selection because they had been deleterious. On the assumption of 60 000 genes, and 240 000 generations since human-chimpanzee divergence, they estimated a deleterious rate of 1.6 mutations per person per generation out of a total of 4.2 mutations per person generation.

The estimated deleterious mutation rates in chimpanzees and gorillas were very similar but those for rodent-specific lineages are about one order of magnitude less, possibly because of much smaller numbers of germ cell divisions in rodents. The very high deleterious rate in hominids may even be an underestimate. If the total gene number were 80 000 and an average coding sequence was 1800 nucleotides, the estimated deleterious mutation rate would be 2.5 mutations per person per generation and, on other grounds too, a more likely rate has been considered to be three deleterious mutations per person per generation.

So with three genetic deaths per person why are we not extinct? The data would suggest that harmful mutations need to be weeded out in clusters at a time. One way to achieve this would be if natural selection operated such that individuals with the most mutations are preferentially eliminated (e.g. harmful mutations interact). This could only happen in a sexual species where mutations are shuffled each generation by genetic recombination, and so the existence of such a high deleterious mutation rate has been taken as further vindication that sex (meiotic recombination) is an efficient way to eliminate harmful mutations.

Pathogenic mutations are preferentially located at certain types of intragenic DNA sequence

Pathogenic mutations can occur at three types of DNA sequence at a gene locus.

The coding sequence of the gene. This is where the great majority of recorded pathogenic mutations have been identified. Those due to nucleotide substitution are, in the vast majority of cases, nonsynonymous substitutions and mostly occur at first and second base positions of codons. However, very rarely, a synonymous codon substitution is not neutral as expected, but may cause disease by activating a cryptic splice site. Because of its relatively high mutability, the CpG dinucleotide is often located at hotspots for pathogenic mutation in coding DNA (Cooper and Youssoufian, 1988). Other hotspots include tandem repeats within coding DNA (see below).

Intragenic noncoding sequences. This is restricted to sequences which are necessary for correct expression of the gene, such as important intronic elements, notably the highly conserved GT and AG dinucleotides at the ends of introns, but also conserved elements of the untranslated sequences. Often such mutations represent a small component (~10–15%) of the total pathogenic mutations at a gene locus (Cooper *et al.*, 1995). However, in some disorders pathogenic splicing mutations may be common. In the case of the collagen disorder osteogenesis imperfecta they constitute a very common pathological mutation which is second in frequency only to substitutions leading to replacement of the highly conserved, structurally important glycine residues. The collagen genes have small exons and a comparatively large number of introns (often more than 50 and as many as 106 in the case of the *COL7A1* gene), making them exceptional targets for splicing mutations. Occasional pathogenic mutations have been recorded in the 5' UTR (such as in the case of hemophilia B Leyden) and appear to exert their effect at the transcriptional level. Several examples are also known of pathogenic mutations in the 3' UTR.

Regulatory sequences outside exons. Most mutations located in regulatory sequences have been identified in conserved elements located just upstream of the first exon, notably promoter elements. In addition, other more distantly located regulatory elements may be sites of pathological mutation. For example, deletions which eliminate the β-globin LCR but leave the β-globin gene and its promoter intact result in almost complete abolition of β-globin gene expression and contribute to β-thalassemia. Clearly, in some cases a gene may be regulated by the product of a distantly related gene. For example, in the case of rare variants of α-thalassemia with mental retardation, the α-globin gene and its promoter may show no evidence of pathological mutation and the disease maps to an X-linked gene which encodes a transcription factor, one of whose target sequences is presumably the α-globin gene.

The mitochondrial genome is a hotspot for pathogenic mutations

Because of the very large size of the human nuclear genome, most mutations occur in nuclear DNA sequences. By comparison, the mitochondrial genome is a small target for mutation (about 1/200 000 of the size of the nuclear genome). Unlike nuclear genes, mitochondrial genes are present in numerous copies (there are thousands of copies of the mtDNA molecule in each human somatic cell; some cells, such as brain and muscle cells, have particularly high oxidative phosphorylation requirements and so more mitochondria). The mtDNA is inherited from the maternal oocyte, which is an exceptional cell with many more mtDNA molecules than somatic cells. Given that a mutation in mitochondrial DNA must arise on a single mtDNA molecule, one might intuitively expect that the chances of a single mtDNA mutation becoming *fixed* would be very low and the mutation rate correspondingly low. On these grounds, one could anticipate that the proportion of clinical disease due to pathogenic mutation in the mitochondrial genome should be extremely low. Instead, the frequency of 'mitochondrial disorders' is rather high and the mitochondrial genome can be considered to be a mutation hotspot. Different factors can explain this apparent paradox:

Differential target size for pathogenic mutation. Pathogenesis is associated with mutations in coding DNA and the mitochondrial genome has a much higher percentage of coding DNA (93%) than found in the nuclear genome (3%). When this is taken into consideration, however, there is still a large imbalance: about 100 Mb of coding DNA in the nuclear genome but only 15.4 kb of coding sequence in the mitochondrial genome, giving a target ratio of 6000:1 in favor of the nuclear genome.

High mutation rate in mtDNA. The mitochondrial genome is much more prone to nucleotide change than the nuclear genome. Even although about 100 000 copies of the mitochondrial genome are maternally inherited in the fertilized oocyte there are mechanisms which permit rapid fixation of mutations in mitochondrial DNA.

The combination of mtDNA instability and a high fixation rate means that the mutation rate in mitochondrial DNA is very high. Mutations have been reported to be fixed in the mitochondrial genomes of animal cells at a rate which is about 10 times greater than occurring in equivalent sequences in the nuclear genome. This means that the small recombination-deficient animal mtDNA molecules appear to be evolving remarkably rapidly, corresponding to about 2–4% sequence divergence per million years. In contrast, plant mtDNA molecules are comparatively large (150 kb-2.5 Mb), have introns, engage in recombination and are evolving comparatively slowly.

The high instability of mtDNA has been postulated to result from several factors. The high rate of production of reactive oxygen intermediates by the respiratory chain is thought to cause substantial oxidative damage to mtDNA which, unlike nuclear DNA, is not protected by histones. The mtDNA also has to undergo many more rounds of replication than chromosomal DNA. Although several well-characterized mtDNA repair systems are now known, some frequent mutations cannot be repaired, including thymidine dimers.

Many different factors govern the expression of pathogenic mutations

The degree to which a pathogenic mutation results in an aberrant phenotype depends on several factors:

The mutation class and the way in which the expression of the mutant gene is altered. This may depend on the location of the mutation within the gene. Many pathogenic mutations result in abolition or substantial reduction of gene expression, but some lead to inappropriate expression. For example, overexpression of a gene product may cause an abnormal phenotype where gene dosage needs to be carefully regulated, and ectopic expression, that is expression in tissues where the gene is not normally expressed, may also be harmful.

The degree to which aspects of the aberrant phenotype are expressed in the heterozygote. The presence of a single normal allele may be sufficient to maintain a clinically normal phenotype (as in recessively inherited disorders), or a milder phenotype when compared with that of mutant homozygotes, as in dominantly inherited disorders where the mutation is a simple *loss of function* mutation.

The degree to which expression of a mutant phenotype is influenced by other gene products. The same mutant allele can have different phenotypic effects on different genetic backgrounds, depending on particular alleles at other gene loci (*modifier genes*).

The proportion and nature of cells in which the mutant gene is present. Generally, mutations which are present in all the cells of an individual (inherited mutations) or in many of them (somatic mutations acquired very early in development) are likely to have a more profound effect than those present in a few cells (somatic mutations which arise at much later stages) or in cell types where the relevant gene is not expressed. Cancers, however, arise from unregulated division of cells produced from a single original mutant cell.

The parental origin of the mutation. This is only known to be important in the case of the few genes which are *imprinted.*

Most splicing mutations alter a conserved sequence required for normal splicing, but some occur in sequences not normally required for splicing

Many genes naturally undergo alternative forms of RNA splicing. In addition, mutations can sometimes produce an aberrant form of RNA splicing which is pathogenic. Sometimes this results in the sequences of whole exons being excluded from the mature RNA (exon skipping) or retention of whole introns. On other occasions, the abnormal splicing pattern may exclude part of a normal exon or result in new exonic sequences. Point mutations which alter a conserved sequence that is normally required for RNA splicing are comparatively common. Occasionally, however, aberrant splicing of a gene can be induced by mutation of other sequence elements which resemble splice donor or splice acceptor sequences but which are not normally involved in splicing (cryptic splice sites).

Mutations which alter important splice site signals: Often such mutations occur at the essentially invariant GT and AG dinucleotides located respectively at the start of an intron (*splice donor*) or at its end (*splice acceptor*). Flanking these important signals, however, are other conserved sequence elements which, if mutated, can also cause aberrant splicing. Mutations which alter such sequences can have different consequences:

Failure of splicing causing intron retention. This can occasionally result, for example, when an intron is small and the neighboring sequence lacks alternative legitimate splice sites or *cryptic splice sites* (sequences which resemble the consensus splice site sequences but which are not normally used by the splicing apparatus). The introduction of intronic sequence into the coding sequence of a mature mRNA will, at the very least, introduce additional amino acids and may cause a frameshift.

Exon skipping. The splicing apparatus uses an *alternative* legitimate splice site. Mutation of a splice donor sequence results in skipping of the upstream exon while mutation of the splice acceptor sequence results in skipping of the downstream exon. Often, the exclusion of an exon has a profound effect on gene expression: it may result in a frameshift, an unstable RNA transcript, or a nonfunctional polypeptide because of a loss of a critical group of amino acids.

In addition to the above, the *branch site* used in splicing may be mutated leading to defective splicing.

Mutations of sequences which are not normally important for RNA splicing: Cryptic splice sites coincidentally resemble the sequences of authentic splice sites but are not normally used in splicing, unless a mutation alters the sequence so that the splicing apparatus now recognizes it as a normal splice site.

Mutations that introduce a premature termination codon usually result in unstable mRNA but other outcomes are possible

Several different classes of mutation can introduce a premature termination codon (chain terminating mutations). Nonsense mutations produce a premature termination codon simply by substituting a normal codon with a stop codon. Frameshifting insertions and deletions usually also introduce a premature termination codon not too far downstream of the mutation site. This happens because there is no selection pressure to avoid stop codons in the other translational reading frames and so given established nucleotide frequencies, at least one stop codon is usually encountered within a stretch of 100 nucleotides downstream of the mutation site. A variety of splice site mutations too can introduce a premature termination codon e.g. by skipping of a single exon containing a number of nucleotides that cannot be divided by 3. There are several possible consequences for gene expression for chain-terminating mutations:

Unstable mRNA. This is by far the most frequent consequence. A mRNA carrying a premature codon is usually rapidly degraded *in vivo* by a form of RNA surveillance known as nonsense-mediated mRNA decay (Hentze and Kulozik, 1999; Culbertson, 1999). This can avoid the potentially lethal consequences of producing a truncated polypeptide which could interfere with vital cell functions.

Truncated polypeptide. The production of a polypeptide truncated at the C terminus is a very rare outcome *in vivo* (the well-known protein truncation test which assays for mutations introducing a premature termination codon is carried out using an *in vitro* transcription-translation system). Nevertheless, some truncated polypeptides are produced *in vivo* (see, for example, Lehrman *et al.*, 1987). The effect on gene expression may be difficult to predict and will depend among other things on the extent of the truncation, the stability of the polypeptide product and its ability to interfere with expression of normal alleles.

Exon skipping. Some nonsense mutations appear to induce skipping of constitutive exons *in vivo*. For example, a nonsense mutation in the middle of exon 51 of the *FBN1* fibrillin gene (corresponding to the C terminus of the protein) causes that exon to be skipped (Dietz *et al.*, 1993). As a result of exon skipping the abnormally spliced mRNA uses the normal stop codon and escapes nonsense-mediated mRNA decay unlike any full length mRNA which may be produced from the pre-mRNA. The abnormally spliced *FBN1* mRNA accumulates and is translated to give a dominant negative protein lacking C-terminal sequences.

The pathogenic potential of repeated sequences

The human genome, like other mammalian genomes, has a very high proportion of DNA sequences that are repeated. Tandem repeats in coding DNA include very short nucleotide repeats, moderately sized repeats and very large repeats that can include whole genes. Depending on the degree of sequence homology between the repeats, tandem repeats are liable to a variety of different genetic mechanisms causing sequence exchange between the repeats. Often such sequence exchanges result in changes in the number of tandem repeats. A reduction in repeat number can often result in a pathogenic deletion, but expansion by sequence duplication can be pathogenic too. Certain chromosomal regions, notably the subtelomeric and pericentromeric regions, harbor large tracts of duplicated DNA and instability of such regions can predispose to disease. Interspersed repeats can also cause pathogenic mutations by a different variety of mechanisms.

Slipped strand mispairing of short tandem repeats predisposes to pathogenic deletions and frameshifting insertions

Insertions and deletions in coding DNA are rare because they usually introduce a translational frameshift. However, occasionally, a series of tandem repeats of a small number of nucleotides occurs by chance in the coding sequence for a polypeptide. Such repeats, like classical microsatellite loci, are comparatively prone to mutation by slipped strand mispairing. As a result, the copy number of tandem repeats is liable to fluctuate, introducing a deletion or an insertion of one or more repeat units. If the mutation occurs in polypeptide-encoding DNA, a resulting deletion will often have a profound effect on gene expression. Frameshifting deletions will normally result in abolition of gene expression. Even if the deletion does not produce a frameshift, deletions of one or more amino acids can still be pathogenic. Small frameshifting insertions will also be expected to lead to loss of gene expression and often the insertion is a tandem repeat of sequences flanking it. However, nonframeshifting insertions would often not be expected to be pathogenic, unless the insertion occurs in a critically important region, destabilizing an essential structure or impeding gene function in some way. Note that large triplet repeat expansions can lead to disease by mechanisms that are not understood at present.

Rapid large-scale expansion of intragenic triplet repeats can cause a variety of diseases but the mutational mechanism is not well understood

Sometimes microsatellites within or in the immediate vicinity of a gene can expand to considerable lengths and affect gene expression, causing disease. In some cases, a modestly expanded repeat which causes disease may be perfectly stable and be propagated without change in size through several generations. For example, triplet repeat expansion leading to long polyalanine tracts in the *HOXD13* gene cause a form of synpolydactyly, probably as a result of unequal crossover (Warren, 1997), but the expanded repeat is stable (Akarsu *et al.*, 1996). In other cases, however, the expanded triplet repeat is *unstable* and the discovery that human disease can be caused by large-scale expansion of highly unstable trinucleotide repeats was quite unexpected. Studies in other organisms had not revealed precedents for such a phenomenon, but the list of human examples is now considerable. In addition to unstable triplet repeat expansion, the majority of disease alleles at the cystatin B gene which cause progressive myoclonus epilepsy involve expansions of a 12 nucleotide repeat (CCCCGCCCCGCG; Lalioti *et al.*, 1998).

Tandem trinucleotide repeats are not infrequent in the human genome. Although there are 64 possible trinucleotide sequences, when allowance is made for cyclic permutations $(CAG)_n = (AGC)_n = (GCA)_n$ and reading from either strand [5' $(CAG)_n$ on one strand = 5' $(CTG)_n$ on the other], there are only 10 different trinucleotide repeats. Most of these are known as usefully polymorphic microsatellite markers but, in addition, certain repeats show anomalous behavior which can cause abnormal gene expression.

In each case, repeats below a certain length are stable in mitosis and meiosis while, above a certain threshold length, the repeats become extremely unstable. These unstable repeats are virtually never transmitted unchanged from parent to child. Both expansions and contractions can occur, but there is a bias towards expansion. The average size change often depends on the sex of the transmitting parent, as well as the length of the repeat. Genes containing unstable expanding trinucleotide repeats fall into two major classes:

Genes which show modest expansions of $(CAG)_n$ repeats within the coding sequence. Typically, the stable and nonpathological alleles have 10–30 repeats, while unstable pathological alleles have modest expansions, often in the range of 40–200 repeats. Transcription and translation of the gene are not affected by the expansion. The resulting protein product shows a gain of function: its long polyglutamine tract causes it to aggregate within certain cells and kill them.

Genes which show very large expansions of a noncoding repeat. For some genes, various types of triplet repeat (e.g. CGG, CCG, CTG, GAA) found in the promoter, the untranslated regions or intronic sequences can undergo very large expansions in such a way as to inhibit gene expression, causing loss of function. Typically, the stable and nonpathological alleles have 5–50 repeats, while unstable pathological alleles have several hundreds or thousands of copies.

Intergenerational changes are normally reported as parent-child comparisons of blood lymphocyte DNA. There is little information about when in gametogenesis, fertilization or embryogenesis the changes arise. Limited studies of sperm show that highly expanded *DM* and *FRAXA* (fragile-X syndrome) repeats are not transmitted by affected males, although modest expansions can be. The largest expansions in Huntington disease (which, however, are small compared with large *FRAXA* or *DM* expansions) are seen in sperm, consistent with the observation that the severest cases inherit the disease from their father. At least for *FRAXA*, *DM* and Kennedy disease, the expanded repeats are mitotically unstable, so that a blood sample shows a smear of heterogeneous expanded repeats sizes. However, *in vitro*, even large repeats are stable. Thus, whatever the mechanism, it is not operative in all cells.

The basis of the unstable expansions is very largely unknown and this type of mutagenic event has not been identified thus far in genetically tractable organisms such as *E. coli*, yeast or *Drosophila*. There is also evidence that the unstable expansion mechanism may not have a parallel in some other mammals such as mice. Human transgenes containing long trinucleotide repeats show virtually no instability after being propagated through several generations in transgenic mice whereas the same sequences may show a 100% probablity of expansion when transmitted in the human germline (Djian, 1998). Investigations have also suggested that arrays of triplet repeats may be able to form alternative DNA structures, such as DNA hairpins, triplex DNA and quadruplex DNA (Sinden, 1999) but their significance if formed *in vivo* is unknown. The repeats have also been envisaged as possible protein-binding sites and protein-binding at the RNA level has also been envisaged to contribute to pathogenesis in some cases, notably in myotonic dystrophy.

Slipped strand mispairing is likely to be a component of the expansion mechanism, given the observation that interrupted repeats appear to be stable and only homogeneous repeats are unstable. For example, in spinocerebellar ataxia type 1, 123/126 normal sized CAG repeats were interrupted by one or two CAT triplets, while 30/30 expanded alleles contained no interruption. One problem with all these mispairing mechanisms is that they should result in contractions as well as expansions and this is not seen. Instead, after a certain threshold size, there appears to be a clear bias towards continued expansion of the size of the repeat unit array. Because understanding of trinucleotide repeats is progressing very rapidly at the time of writing, the reader is advised to consult a recent review for more information.

Tandemly repeated and clustered gene families may be prone to pathogenic unequal crossover and gene conversion-like events

Many human and mammalian gene clusters contain nonfunctional pseudogenes which may be closely related to functional gene members. Interlocus sequence exchanges between pseudogenes and functional genes can result in disease by removing or altering some or all of the sequence of a functional gene. For example, unequal crossover (or unequal sister chromatid exchange) between a functional gene and a related pseudogene can result in deletion of the functional gene or the formation of fusion genes containing a segment derived from the pseudogene. Alternatively, the pseudogene can act as a donor sequence in gene conversion events and introduce deleterious mutations into the functional gene.

The classical example of pathogenesis due to gene-pseudogene exchanges is steroid 21-hydroxylase deficiency, where over 95% of pathogenic mutations arise as a result of sequence exchanges between the functional 21-hydroxylase gene, *CYP21B*, and a very closely related pseudogene, *CYP21A*. The two genes occur on tandemly repeated DNA segments approximately 30 kb long which also contain other duplicated genes, notably the complement C4 genes, *C4A* and *C4B*. Large pathogenic deletions uniformly result in removal of about 30 kb of DNA, corresponding to one repeat unit length, and analysis of *de novo* 21-hydroxylase deficiency mutations has provided strong evidence for pathogenic deletions arising as a result of meiotic unequal crossover.

Virtually all of the 75% of pathogenic point mutations are copied from deleterious mutations in the pseudogene, suggesting a gene conversion mechanism. Analysis of one such mutation which arose *de novo* suggests that the conversion tract is a maximum of 390 bp. Gene conversion events are also found in the duplicated C4 genes, both of which are normally expressed. A likely priming event for conversions in the *CYP21-C4* gene cluster is unequal pairing of chromatids so that a *CYP21A-C4A* unit pairs with a *CYP21B-C4B* unit.

Interspersed repeats often predispose to large deletions and duplications

Short direct repeats: In several cases, the endpoints of deletions are marked by very short direct repeats. For example, the breakpoints in numerous pathogenic deletions in mtDNA occur at perfect or almost perfect short direct repeats. Of these, the most common is a deletion of 4977 bp which has been found in multiple patients with Kearns-Sayre syndrome, an encephalomyopathy characterized by external ophthalmoplegia, ptosis, ataxia and cataract. The deletion results in elimination of the intervening sequence between two perfect 13 bp repeats and loss of the sequence of one of the repeats. The mitochondrial genome is recombination-deficient and Shoffner *et al.* 1989 have postulated that such deletions arise by a replication slippage mechanism, similar to that occurring at short tandem repeats. Partial duplications of the mitochondrial genome are also distinctive features of certain diseases, notably Kearns-Sayre syndrome. The ends of the duplicated sequences, like those of the common deletions, are often marked by short direct repeats, and the mechanisms of duplication and deletion appear to be closely related.

The Alu repeat as a recombination hotspot: Some large-scale deletions and insertions may be generated by pairing of nonallelic interspersed repeats, followed by breakage and rejoining of chromatid fragments.

For example, the *Alu* repeat occurs approximately once every 4 kb and mispairing between such repeats has been suggested to be a frequent cause of deletions and duplications. Some large genes have many internal *Alu* sequences in their introns or untranslated sequences, making them liable to frequent internal deletions and duplications.

For example, the 45-kb low density lipoprotein receptor gene has a relatively high density of *Alu* repeats (approximately one every 1.6 kb). A very high frequency of pathogenic deletions in this gene are likely to involve an *Alu* repeat, usually at both endpoints, and occasional pathogenic intragenic duplications also involve *Alu* repeats (Hobbs *et al.*, 1990).

Such observations have suggested a general role for *Alu* sequences in promoting recombination and recombination-like events. Initial gene duplications in the evolution of clustered multigene families may often have involved an unequal crossover event between *Alu* repeats or other dispersed repetitive elements. It should be noted, however, that some *Alu*-rich genes do not appear to be loci for frequent *Alu*-mediated recombination.

Pathogenic inversions can be produced by intrachromatid recombination between inverted repeats

Occasionally, clustered inverted repeats with a high degree of sequence identity may be located within or close to a gene. The high degree of sequence similarity between inverted repeats may predispose to pairing of the repeats by a mechanism that involves a chromatid bending back upon itself. Subsequent chromatid breakage at the mispaired repeats and rejoining can then result in an inversion, in much the same way as the natural mechanism used for the production of some immunoglobulin ê light chains.

The classic example of pathogenic inversions is a mutation which accounts for more than 40% of cases of severe hemophilia

A. Intron 22 of the factor VIII gene, *F8*, contains a CpG island from which two internal genes are transcribed: *F8A* in the opposite direction to the host gene *F8*, and *F8B* in the same direction as *F8*.

F8A belongs to a gene family with two other closely related members located several hundred kilobases upstream of *F8* gene and transcribed in the opposite direction to *F8A*. As a result, the region between the *F8A* gene and the other two members is susceptible to inversions - the *F8A* gene can pair with either of the other two members on the same chromatid, and subsequent chromatid breakage and rejoining in the region of the paired repeats results in an inversion which disrupts the factor VIII gene.

DNA sequence transposition is not uncommon and can cause disease

Defective gene expression due to DNA transposition is comparatively rare and represents only a small component of molecular pathology.

However, several examples have been recorded of genetic deficiency due to insertional inactivation by retrotransposons. For example, in one study, hemophilia A was found to arise in two out of 140 unrelated patients as a result of a *de novo* insertion of a LINE-1 (*Kpn*) repeat into an exon of the factor VIII gene (Kazazian *et al.*, 1988).

Other instances are known of insertional inactivation by an actively transposing *Alu* element, as in a case of neurofibromatosis type 1 (Wallace *et al.*, 1991). Additionally, a number of other examples have been recorded of pathogenesis due to intragenic insertion of undefined DNA sequences.

DNA repair

DNA in cells suffers a wide range of damage:

- purine bases are lost by spontaneous fission of the base-sugar link;
- cytosines, and occasionally adenines, spontaneously deaminate to produce uracil and hypoxanthine respectively;

- many chemicals, for example alkylating agents, form adducts with DNA bases;
- reactive oxygen species in the cell attack purine and pyrimidine rings
- ultraviolet light causes adjacent thymines to form a stable chemical dimer;
- mistakes in DNA replication result in incorporation of a mismatched base;
- ionizing radiation causes single- or double-strand breaks;
- mistakes in replication or recombination leave strand breaks in DNA.

All these lesions must be repaired if the cell is to survive. The importance of effective DNA repair systems is highlighted by the severe diseases affecting people with deficient repair systems.

DNA repair usually involves cutting out and resynthesizing a whole area of DNA surrounding the damage

To cope with all these forms of damage, cells must be capable of several different types of DNA repair. DNA repair seldom involves simply undoing the change that caused the damage. Almost always a stretch of DNA containing the damaged nucleotide(s) is excised and the gap filled by resynthesis. There are at least five main types of DNA repair in human cells:

Direct repair reverses the DNA damage. A specific enzyme is able to dealkylate O^6-alkyl guanine directly. In bacteria thymine dimers can be removed in a photoreactivation reaction that depends on visible light and an enzyme, photolyase. Mammals possess enzymes related to photolyase, but use them for a quite different purpose, to control their circadian clock.

Base excision repair (BER) uses glycosidase enzymes to remove abnormal bases. An endonuclease, AP endonuclease, cuts the sugar-phosphate backbone at the position of the missing base. A few nucleotides of the DNA strand are stripped back by exonucleases, the gap is filled by resynthesis, and the remaining nick is sealed by DNA ligase III. The same process is used to repair spontaneous depurination. Interestingly, no human diseases caused by defective BER are known. Maybe any such defect would be lethal, since BER corrects much the commonest type of DNA damage.

Nucleotide excision repair (NER) removes thymine dimers and large chemical adducts. *Figure 9.21* illustrates the process. Defects in NER cause the autosomal recessive disease xeroderma pigmentosum (Lambert *et al.*, 1998). Seven complementation groups, XPA-XPG, have been defined by cell fusion studies. XP patients are exceedingly sensitive to UV light. Sun-exposed skin develops thousands of freckles, many of which progress to skin cancer.

Post-replication repair is required to correct double-strand breaks (Haber, 1999). The usual mechanism is a gene conversion-like process (recombinational repair), where a single strand from the homologous chromosome invades the damaged DNA helix.

Alternatively, broken ends are rejoined regardless of their sequence, a desperate measure that is likely to cause mutations. The eukaryotic machinery for recombination repair is less well defined than the excision repair systems.

Mismatch repair corrects mismatched base pairs caused by mistakes in DNA replication. Cells deficient in mismatch repair have mutation rates 100–1000 times higher than normal, with a particular tendency to replication slippage in homopolymeric runs. In humans the mechanism involves at least five proteins and defects cause hereditary nonpolyposis colon cancer.

All these systems, except for direct repair, require exo- and endonucleases, helicases, polymerases and ligases, usually acting in multiprotein complexes that have some components in common. Sorting out the individual pathways has been greatly aided by the very strong conservation of repair mechanisms across the whole spectrum of life.

Not only the reaction mechanisms but also the protein structures and even gene sequences are often conserved from

E. coli to man. A downside of the conservation is a confusing gene nomenclature, referring sometimes to human diseases (*XPA* etc.), sometimes to yeast mutants (*RAD* genes) and sometimes to mammalian cell complementation systems (ERCC = excision repair cross-complementing) - for example *XPD*, *ERCC2* and *RAD3* are the same gene in man, mouse and yeast.

Generally eukaryotes have multiple proteins corresponding to each single protein in *E. coli*, so that, for example, nucleotide excision repair requires six proteins in *E. coli* but at least 30 in mammals.

DNA repair systems share components and processes with the transcription and recombination machinery

As well as sharing components with each other, many repair systems share components with the machinery for DNA replication, transcription and recombination. DNA polymerases and ligase are required for both DNA replication and resynthesis after excision of a defect.

The recombination machinery is involved in double-strand break repair. The link with transcription is particularly intriguing (Lehmann, 1995). The general transcription factor TFIIH is a multiprotein complex that includes the XPB and XPD proteins. TFIIH exists in two forms.

One form is concerned with general transcription and the other with repair, probably specifically repair of transcriptionally active DNA. This system is deficient in two rare diseases, Cockayne syndrome (MIM 216400) and trichothiodystrophy (MIM 601675). Clinically and in cell biology, CS and TTD both overlap XP, and in some cases the same genes are responsible, but CS and TTD patients have developmental defects that presumably reflect defective transcription, and they do not have the cancer susceptibility of XP patients.

Hypersensitivity to agents that damage DNA is often the result of an impaired cellular response to DNA damage, rather than defective DNA repair

Many human diseases that involve hypersensitivity to DNA-damaging agents, or a high level of cellular DNA damage, are not caused by defects in the DNA repair systems themselves, but by a defective cellular response to DNA damage. Normal cells react to DNA damage by stalling progress through the cell cycle at a checkpoint until the damage has been repaired, or triggering apoptosis if the damage is irrepairable.

Part of the machinery for doing this involves the ATM protein. Briefly, it senses DNA damage and relays the signal to the p53 protein, the 'guardian of the genome'. People with no functional ATM have ataxia telangiectasia.

Their cells are hypersensitive to radiation, and they have chromosomal instability and a high risk of malignancy, but the DNA repair machinery itself is intact. Fanconi anemia (MIM 227650) is another heterogeneous group of diseases (at least five complementation groups) marked by defective responses to DNA damage, without specific defects in DNA repair.

7

PHYSICAL AND TRANSCRIPT MAPPING

A wide variety of physical mapping strategies have been used to analyze the DNA of complex eukaryotic genomes. In the following sections, an arbitrary distinction is made between two classes of physical mapping:

- low resolution physical mapping - the smallest map unit that can be resolved is typically one to several megabases of DNA;
- high resolution physical mapping - the resolution is typically very high, from hundreds of kilobases to a single nucleotide.

Since mammalian DNA has only a very small percentage of coding DNA (~3% in the case of the human genome), a variety of transcript mapping methods have been developed for selectively identifying and studying transcribed sequences, the most important and, arguably, most interesting component of the genome.

Low resolution physical mapping

Somatic cell hybrid panels can permit chromosomal localization of any human DNA sequence

Under certain experimental conditions, cultured cells from different species can be induced to fuse together, thereby generating somatic cell hybrids. In human genetic mapping, hybrid cells are typically constructed by fusing human cells and mouse or hamster cells. The initial fusion products are described as heterokaryons because the cells contain both a human and a rodent nucleus. Eventually, heterokaryons proceed to mitosis, and the two nuclear envelopes dissolve. Thereafter, the human and rodent chromosomes are brought together in a single nucleus.

The hybrid cells are unstable. For reasons that remain unknown, most human chromosomes fail to replicate in subsequent rounds of cell division and are lost. This gives rise, eventually, to a variety of more-or-less stable hybrid cell lines, each with the full set of rodent chromosomes plus a few human chromosomes. The loss of the human chromosomes occurs essentially at random, but can be controlled by selection.

The human chromosomes in somatic cell hybrids can conveniently be identified by PCR screening with sets of chromosome-specific primers.

By collecting hybrid cell lines with different human chromosome contents it is possible to generate a hybrid cell panel that can be used to map any human DNA sequence to a specific chromosome.

To do this, each of the hybrid cell lines is tested for the presence of the human sequence of interest. A PCR assay can be used with primers specific for that sequence (this is a form of STS content mapping) or the relevant DNA sequence can be labeled and used as a hybridization probe. Both of these methods require that the human sequence should be distinguishable from any homologous rodent sequence.

Somatic cell hybrids could also be used to establish which chromosome encodes a human enzyme or other gene product by biochemical typing of the hybrids. Whichever human chromosome contains the sequence of interest, it will be present in all the cell lines that are positive for the test and absent from all cell lines that are negative for the test.

Monochromosomal hybrids generated by microcell fusion allow direct and rapid chromosomal localization of human DNA clones: Traditional somatic cell hybrids generally contain several human chromosomes. Microcell fusion can be used to generate monochromosomal hybrids that contain only a single human chromosome. The donor human cells are subjected to prolonged mitotic arrest by continued exposure to an inhibitor of mitotic spindle formation, such as colcemid. This causes the chromosome content of a cell to become partitioned into discrete subnuclear packets (micronuclei). Centrifugation in the presence of cytochalasin B (a mitotic spindle inhibitor), results in the formation of microcells. Microcells consist of a single micronucleus with a thin rim of cytoplasm, surrounded by an intact plasma membrane. In much the same way as for normal donor cells, the microcells can be fused with recipient rodent cells to generate hybrids (microcell-mediated chromosome transfer). Some microcell hybrids contain a few donor chromosomes, but the simplest contain a single donor chromosome (see, for example, Warburton *et al.*, 1990).

Monochromosomal human-rodent hybrid cell lines have been established for each of the human chromosomes (Cuthbert *et al.*, 1995). If the sequence of a particular human DNA is known, a PCR assay can be established and used to type a complete monochromosomal hybrid panel, thereby establishing the chromosomal location quickly. Alternatively, if no DNA sequence is available, a human fragment of interest can still be labeled and used as a hybridization probe against a Southern blot of genomic DNA samples from a monochromosomal hybrid panel.

Subchromosomal mapping is possible using hybrid cells containing defined portions of a human chromosome

Conventional somatic cell hybrids are a relatively crude tool for physical mapping. More refined mapping is possible using hybrids that contain only part of a particular human chromosome. Translocation hybrids and deletion hybrids are made using donor human cells that have a chromosomal translocation or deletion. To be useful, the hybrids must lack the normal homolog of the chromosome of interest. Such hybrids can be used for subchromosomal mapping of a human sequence-tagged site or biochemical marker. They are especially useful for defining the sequences removed by microdeletions, by segregating the deletion-carrying chromosome away from its normal homolog.

Human DNA clones can be physically mapped to any desired resolution by STS mapping in panels of hybrid cells that contain defined fragments of the human genome

A more general approach to producing tools for physical mapping involves artificially breaking human chromosomes and transferring the fragments into rodent cells.

Chromosome-mediated gene transfer: One of the first techniques to use this approach was chromosome-mediated gene transfer (CMGT). Fragments of purified mitotic chromosomes from a donor, such as a human fibroblast, are coprecipitated with calcium phosphate on to the surface of a recipient rodent cell line in monolayer culture. Human chromosome fragments enter the recipient cells, such as mouse fibroblasts, and integrate into the chromosomes, resulting in stable transformation. As a result, hybrids can be established that retain segments of human DNA (transgenomes) of a size that is useful for mapping (usually in the range of 1–50 Mb). However, the transgenomes are prone to frequent rearrangements, so CMGT is more suited to functional assays of complex loci than as a mapping tool.

Irradiation fusion gene transfer: The most valuable hybrids for gene mapping are radiation hybrids (see Walter *et al.*, 1994). Donor cells are subjected to a lethal dose of radiation which fragments their chromosomes. The average size of a fragment is a function of the dose of radiation. After irradiation the donor cells are fused with recipient cells of a different species. A selection system is used to pick out recipient cells that have taken up some of the donor chromosome fragments. These cells are useful for mapping insofar as they have taken up a random set of other chromosome fragments from the donor, as well as the selected fragment. Stably incorporated donor fragments are either integrated into rodent chromosomes or are assembled into novel human minichromosomes formed around fragments containing a functional centromere.

Although this procedure was first proposed by Goss and Harris in 1975, it was not used seriously until 1990, when hybrids were constructed using irradiated mono-chromosomal hybrid cells as donors (Cox *et al.*, 1990). When a set of DNA markers from the human chromosome is assayed in a panel of such radiation hybrids, the patterns of cross-reactivity can be used to construct a map. The principle is very similar to meiotic linkage analysis as described in Chapter 11: the nearer together two DNA sequences are on a chromosome, the lower the probability that they will be separated by the chance occurrence of a breakpoint between them.

The frequency of breakage between two markers can be defined by a value θ, analogous to the recombination frequency in meiotic mapping. θ varies from 0 (the two markers are never separated) to 1.0 (the two markers are always broken apart). As in meiotic mapping, θ underestimates the distance between markers that are far apart on the same chromosome, in this case because a cell can take up two markers on separate fragments. A more accurate estimate is provided by a mapping function, $D = -\ln(1 - \theta)$, which is analogous to the Haldane mapping function used in meiotic linkage analysis. D is measured in centiRays (cR). D is dependent on the dosage of radiation, so it is referenced against the number of rads. For example, a distance of 1 cR_{8000} between two markers represents a 1% frequency of breakage between them after exposure to 8000 rad of X-rays.

Radiation hybrids derived from monochromosomal hybrid donor cells have been superseded by whole-genome radiation hybrids where the donor is an irradiated normal human diploid cell. The first such panel consisted of 199 hybrids made by fusing an irradiated 46,XY human fibroblast cell line to TK^- hamster cells (Walter *et al.*, 1994). Gyapay *et al.* (1996) used 404 microsatellite markers of known location to show that this hybrid panel could generate accurate maps, and then used it to map 374 unmapped ESTs. A subset of 93 of the hybrids has been made widely available as the Genebridge 4 panel. The 93 hybrids average 3'% retention of any particular human sequence, with an average fragment size of 25 Mb. Laboratories can map any unknown STS by scoring the 93 Genebridge hybrids and comparing the pattern with patterns of previously mapped markers held on a central server.

This has turned into an extremely powerful and convenient tool for physically mapping any STS or EST. A second human-hamster panel, Stanford G3, was made using a higher dose of radiation, so that the average human fragment size is smaller. The 83 hybrids in G3 average 16% retention of the human genome, with an average fragment size of 2.4 Mb. Thus G3 can be used for finer mapping.

Chromosomal in situ hybridization has been revolutionized by fluorescence in situ hybridization techniques

Chromosomal *in situ* hybridization typically uses an air-dried microscope slide preparation of metaphase chromosomes, in which the chromosomal DNA has been denatured by exposure to formamide before hybridization with a suitable probe. Isotopes such as ^{32}P are not suitable for probe labeling because their high energy causes scattering of the signal. Instead, isotopes with weaker emission of beta radiation have been used. Owing to the high background noise when using such isotopes, however, the genuine signal is not easily identified and complex statistical analysis is required to differentiate genuine signal from background noise.

More recently, the sensitivity and resolution of *in situ* hybridization has been increased significantly by the development of fluorescence*in situ*hybridization (FISH) (Trask, 1991; van Ommen *et al.*, 1995). In this technique, the DNA probe is either labeled directly by incorporation of a fluorescent-labeled nucleotide precursor, or indirectly by incorporation of a nucleotide containing a reporter molecule (such as biotin or digoxigenin) which after incorporation into the DNA is then bound by fluorescently labeled affinity molecule. To increase the intensity of the hybridization signal, large DNA probes are preferred, usually cosmid clones containing around 40 kb of insert. Because such large sequences will contain many interspersed repetitive DNA sequences, it is necessary to use chromosome *in situ* suppression hybridization (Lichter *et al.*, 1990). Essentially, this is a form of competition hybridization: before the main hybridization, the probe is mixed with a large excess of unlabeled total genomic DNA and denatured, thereby saturating the repetitive elements in the probe, so that they no longer mask the signal generated by the unique sequences.

FISH has the advantage of providing rapid results which can be scored conveniently by eye using a fluorescence microscope. In metaphase spreads, positive signals show as double spots, corresponding to probe hybridized to both sister chromatids. Using sophisticated image processing equipment and reporter-binding molecules carrying different fluorophores, it is possible to map and order several DNA clones simultaneously.

The maximum resolution of conventional FISH on metaphase chromosomes is several megabases. The use of the more extended prometaphase chromosomes can permit 1 Mb resolution but, because of problems with chromatin folding, two differentially labeled probe signals may appear to be side-by-side, unless they are separated by distances greater than 2 Mb. Recently, however, new variations have been developed, permitting very high resolution. There are numerous applications for chromosome FISH but a special one is chromosome painting which has permitted a form of molecular karyotyping.

Flow cytometry permits sorting of individual chromosomes and the construction of chromosome-specific DNA libraries

Human chromosomes show considerable variation in size and DNA content. In addition, the base composition can vary considerably: chromosomes that are known to be gene-rich or gene-poor have a comparatively high % (G + C) or % (A + T) respectively. For example, chromosomes 21 and 22 are similar in size and in DNA content, but chromosome 21 is gene-poor and has a relatively high % (A + T), whereas chromosome 22 is gene-rich and has a relatively high % (G + C). As a result of the size differences and differences in base composition, human chromosomes can be separated by flow sorting (also called flow cytometry; see Bartholdi *et al.*, 1987). In this technique, chromosome preparations are stained with a DNA-binding dye which can fluoresce in a laser beam. The amount of fluorescence exhibited by a given chromosome is proportional to the amount of dye bound, which in turn is largely proportional to the amount of DNA, and hence the size of the chromosome. Additionally, certain dyes, such as chromomycin A_3, show preferential binding to GC-rich sequences, while others, such as Hoechst 33258, bind preferentially to AT-rich regions.

Because individual chromosomes can bind different amounts of fluorescent dyes, they can be fractionated in a flow cytometer. A stream of droplets containing stained chromosomes is passed through a finely focused laser beam at a rate of about 1000–2000 chromosomes per second, and the fluorescence of individual chromosomes present in individual droplets is monitored by a photomultiplier. The intensity of fluorescence of the different chromosomes can then be recorded. By using sort windows to define a range of fluorescence intensities corresponding to a particular type of chromosome, droplets containing the desired chromosome can be given an electric charge and deflected from the main stream on to a separate collecting grid.

The resulting flow karyogram may show good separation of some human chromosomes; others such as chromosomes 9–12 are not well separated. However, monochromosomal somatic cell hybrids can be used instead, and often result in good separation of the human chromosome. As a result, panels of chromosome-specific DNA representing each of the 24 human chromosomes have been obtained.

These can be used to obtain a chromosomal localization for any human DNA fragment: the fragment is labeled and hybridized to dot-blots of DNA from the 24 chromosomes ('flow-blots'), or a PCR assay can be used, if appropriate. Additionally, purified DNA from a specific type of chromosome can be amplified by cell-based or PCR-based DNA cloning to generate chromosome-specific DNA libraries (Davies et al., 1981). Chromosome libraries constructed using cosmid vectors have been particularly useful because YACs from a specific chromosome can be related quickly to cosmid clones from the same region. In addition, it has been possible in some cases to construct human chromosome-specific YAC libraries (e.g. McCormick et al., 1993).

High resolution physical mapping: chromatin and DNA fiber FISH and restriction mapping

The physical mapping methods described above typically have a lower resolution limit of one to several megabases; they are complemented by molecular mapping methods which can map DNA in the range 1 bp to several megabases. DNA sequencing provides the ultimate physical map by determining the linear order of single nucleotides, but mapping large DNA regions by sequencing is technically arduous. Two major techniques provide map resolution of less than 1 Mb: restriction mapping and high resolution FISH on naturally extended chromosomes or artificially extended chromatin or DNA fibers.

Very high resolution FISH mapping can be achieved by hybridizing probes to extended chromosomes or artificially extended chromatin or DNA fibers

The linear length of DNA in an average sized human chromosome is about 5 cm, but it is compacted by various hierarchies of folding in metaphase chromosomes to a few microns. As a result, standard *in situ* hybridization against the highly condensed chromatin of metaphase and prometaphase chromosomes does not have a high mapping resolution, with typical upper limits of about one to several megabases of DNA. To obtain higher resolution, DNA probes are hybridized to the naturally extended interphase chromosomes or to artificially stretched chromatin or DNA fibers prepared by a variety of different methods.

In situ hybridization using conventional interphase chromosomes: The chromosomes of interphase nuclei are much more extended than metaphase or prometaphase chromosomes. As a result, FISH analyses of the chromosomes of interphase nuclei can permit a high mapping resolution, and can help to determine the physical mapping order of some syntenic DNA clones. Because the extended chromosomes may loop back at certain regions, an accurate linear order for syntenic DNA clones requires a statistical analysis on the mapping results from many individual interphase nuclei. Even so, the method is best suited to ordering sequences which are separated by intervals within the 50–500 kb range.

FISH mapping on artificially extended chromatin or DNA fibers: High resolution FISH can also be conducted using procedures which cause the DNA of chromosomes on a microscope slide to be extended prior to hybridization. One such method, DIRVISH (direct visual hybridization), involves lysing cells with detergent at one end of a glass slide, tipping the slide, and allowing the DNA in solution to stream down the slide. Such preparations permit extremely high mapping resolutions: from over 700 kb to under 5 kb. More recently, the principle of artificially stretching chromatin fibers has been extended to protein-free DNA: the target DNA is prepared from cells embedded in pulsed field gel electrophoresis blocks. This DNA fiber FISH method used unfixed linearized DNA fibers on a microscope slide and has a resolution of from 500 kb to a few kilobases (see Heiskanen *et al.*, 1996 for a review on high resolution FISH mapping).

Long-range restriction mapping requires enzymes that cut DNA infrequently and fractionation of large restriction fragments by pulsed-field gel electrophoresis

Restriction mapping permits molecular mapping with resolutions which depend on the frequency of the recognition site. Most restriction enzymes which recognize a 4 or 6 bp sequence typically cut vertebrate DNA once every few hundred or few thousand base pairs.

The recognition sequences for rare-cutter restriction nucleases are typically 6–8 bp long and contain one or more CpG dinucleotides which are rare in vertebrate DNA. As a result, they generate fragments that are typically several hundred kilobases in size. Small restriction fragments can be size fractionated conveniently by agarose gel electrophoresis. However, the ability of conventional agarose gel electrophoresis to fractionate large DNA fragments is very limited. The method relies on a *sieving* effect: DNA molecules pass through pores in the agarose gel and small molecules are able to migrate more quickly through the uniform size pores. Above a certain size of DNA fragment, however, the sieving effect is no longer effective and the resolution of DNA fragments above 40 kb is extremely limited.

Large-scale restriction mapping of human DNA can be done directly on the large inserts of YACs or by means of Southern blot hybridization of genomic DNA. Conventionally prepared genomic DNA is not a suitable target for large-scale restriction mapping, because the procedures involved in lysing the cells and purifying the DNA result in shear forces causing considerable fragmentation of the DNA. Instead, the DNA is isolated in such a way as to minimize artificial breakage of the large molecules prior to digesting with restriction endonucleases. To prepare high molecular weight DNA, samples of cells, for example white blood cells, are mixed with molten agarose and then transferred into wells in a block-former and allowed to cool. As a result, the cells become entrapped in solid agarose blocks. The agarose blocks are removed and incubated with hydrolytic enzymes which diffuse through the small pores in the agarose and digest cellular components, but leave the high molecular weight chromosomal DNA virtually intact. Individual blocks containing purified high molecular weight DNA can then be incubated in a buffer containing a rare-cutter restriction endonuclease.

In order to separate large restriction fragments, pulsed-field gel electrophoresis (PFGE), a modified form of agarose gel electrophoresis, is used. Agarose blocks containing the large DNA fragments to be separated are placed in wells at one end of an agarose gel and the DNA migrates in the electric field. However, during a PFGE run, the relative orientation of the gel and the electric field is periodically altered, typically by setting a switch to deliver brief pulses of power, alternatively activating two differently oriented fields. Variants of the technique use a single electric field but with periodic reversals of the polarity (*field inversion gel electrophoresis*), or periodic rotation of the gel or electrodes. In common to each variation is the principle of a discontinuous electric field so that the DNA molecules are intermittently forced to change their conformation and direction of migration during their passage through the gel. The time taken for a DNA molecule to alter its conformation and reorient itself in the direction of the new electric field is strictly size dependent; as a result, DNA fragments up to several megabases in size can be fractionated efficiently.

Assembly of clone contigs

A primary goal of physical mapping is to assemble comprehensive series of DNA clones with overlapping inserts (clone contigs)

The construction of the ultimate physical map (the complete nucleotide sequence) requires considerable time and effort in the case of a very large DNA molecule such as that found in a chromosome. In order to provide a framework for this to be done efficiently, a series of cloned DNA fragments need to be assembled which collectively provide full representation of the sequence of interest. To ensure that there is complete representation, and no gaps, the series of clones should contain overlapping inserts forming a comprehensive clone contig. In principle, contig assembly is facilitated by the way in which genomic DNA libraries are constructed: as part of the strategy for maximizing the representation of a library, the genomic DNA is deliberately subjected to partial digestion with a restriction endonuclease (by reducing the time of incubation and by using low concentrations of enzyme). As a result, individual genomic DNA clones usually contain DNA sequences that partially overlap with the insert DNA of at least some other clones in the library. The cloning step means that the individual DNA fragments are sorted into different cells and so the original positional information of the fragments (how they were related to each other on the original chromosomes) is lost. However, such information can be retrieved by a variety of methods which can identify clones with overlapping inserts.

Chromosome walking means establishing clone contigs from fixed starting points

One widely used technique for identifying clones with overlapping inserts is to use a specific DNA probe from one clone to screen a DNA library. The positively hybridizing clones should contain a DNA sequence that is closely related to the probe, including clones which contain sequences which partly overlap that found in the probe. This has often involved the preparation of a so-called end probe from the starting DNA clone: a fragment located at one end of the insert DNA and preferably present as single copy DNA, is purified and labeled.

Positively hybridizing clones can then be purified and new, distal end probes can be prepared for further rounds of hybridization screening of the DNA library. In the case of genomic DNA libraries, this permits the assembly of a clone contig by bidirectional chromosome walking from a fixed starting point. In the case of mammalian genomes, chromosome walking has frequently involved screening cosmid libraries (cosmid walking). Modern cosmid walking has been simplified enormously by avoiding the need to prepare end probes. Instead, the entire insert can be labeled and used as a probe under conditions which suppress the hybridization signal from DNA sequences in the probe which are highly repetitive in the human genome. Nevertheless, the signal from some other repetitive sequences may be difficult to suppress, causing occasional difficulties with this approach.

Chromosome walking with YACs often involves PCR-based library screening

Chromosome walking by whole clone-clone hybridization is not practically feasible with mammalian YACs: the large amount of repetitive DNA in the inserts means that blocking of the repetitive DNA signal during hybridization (see *Box 5.3*) is technically difficult. Instead, techniques are used to recover short end fragments from individual YACs. The YAC DNA is cleaved with a restriction enzyme which is known to cleave the YAC vector sequence and which cuts frequently in human genomic DNA. Among the cleavage products there will be end fragments containing both the unknown terminal sequence from the insert DNA and the adjacent known vector sequence. Such sequences can then be amplified using various PCR-based methods whereby a primer for a characterized sequence (in this case the sequence of the YAC vector adjacent to the cloning site) is used to permit access to an adjacent uncharacterized sequence (in this case the sequences at the ends of the insert). Frequently used methods include the following:

Inverse-PCR. This is a general method for amplifying DNA flanking a previously characterized region (Ochman *et al.*, 1988). It utilizes primers derived from the extremities of a characterized region which are divergent: the 5' → 3' direction of the two primers point away from each other instead of towards each other. Normally, PCR amplification would be impossible with such primers. However, inverse PCR involves first cutting the DNA so that the desired region is present on a small restriction fragment and then encouraging the DNA to circularize (by conducting the ligations at very low DNA concentrations, the two ends of a single molecule are more likely to come into contact than those on different molecules). Amplification of the uncharacterized flanking regions is then possible using standard conditions.

Vectorette PCR (also called bubble-linker PCR) is a type of anchored PCR which uses a specially designed double-stranded oligonucleotide linker ligated to the target DNA. The double-stranded linker does not contain perfectly complementary oligonucleotides (as is normal, except for overhangs designed to help ligation to target DNA). Instead, the two oligonucleotides of the linker are designed to be complementary at the ends, but not in the middle creating a bubble-like formation. Because the two sequences in the region of the bubble are noncomplementary, it is possible to design a linker strand-specific primer which, with a convergent target DNA-specific primer, permits amplification of the uncharacterized sequence flanking one side of the characterized DNA.

Once retrieved, YAC insert end fragments can be used as hybridization probes to screen colony filters from a YAC library. A widely used alternative, however, is to sequence an end fragment and then design oligonucleotide primers to permit a specific PCR assay for this sequence.

The resulting PCR assay can be used to screen YAC libraries which have been distributed as hierarchical sets of agarose plugs each containing pools of YAC clones. Positive results will identify YACs with overlapping inserts from which new end fragments can be rescued to continue the YAC walk. As, however, YAC clones often have chimeric inserts (composed of two or more fragments that are derived from noncontiguous portions of the genome), the chromosomal location of any new YACs identified in a YAC walk ought to be verified by FISH analyses.

Clone contigs can be assembled rapidly over large fractions of a genome by random clone fingerprinting

Chromosome walking is a highly directional and location-restricted procedure for generating clone contigs: a great deal of effort is invested in generating a contig from a fixed starting point on a specific chromosome, often as a prelude to positional cloning. In order to build clone contigs spanning large amounts of a chromosome, or even a whole genome, a more general, labor-efficient mapping approach is needed. To meet this need, a variety of clone fingerprinting techniques have been used to type clones at random and then integrate the information in order to identify clone contigs over large regions of a genome. For example, this approach has been used to great effect in assembling YAC contigs over significant amounts of the human genome. Clone fingerprinting techniques should be simple and rapid, and the methods described below have been widely used.

Repetitive DNA fingerprinting: Mammalian DNA has a very high component of highly repeated, interspersed DNA sequences such as the human *Alu* and LINE-1 (*Kpn*) repeats (occurring on average about once every 3 kb and 10 kb respectively). Because the spacing between such interspersed repetitive DNA elements will vary in different genomic regions, clones which contain overlapping DNA sequences may be identified by similar patterns following hybridization of repetitive DNA probes to Southern blots of the clone DNA digested with a suitable restriction endonuclease. For example, a whole genome clone fingerprinting approach has been applied to mapping the human genome largely on the basis of repetitive DNA fingerprinting of YACs.

Sequence-tagged site content mapping: If a small amount of DNA sequence (e.g. a few hundred nucleotides) is available for a specific DNA clone, a PCR assay can be developed which is specific for that sequence. The DNA sequence is examined to identify short regions from which oligonucleotide primers can be designed for use in a standard PCR reaction. The site on the original genomic DNA from which the sequence is derived can then be thought of as being tagged by the ability to assay for that sequence. A site for which a specific PCR assay is available is therefore described as a sequence-tagged site (STS). If numerous individual STS sequences have been positioned on a chromosome or subchromosomal area, and YACs have been mapped to the same general area (e.g. by FISH), the ability to type rapidly for STSs allows a rapid method for identifying YACs with overlapping sequences.

Hybridization using Alu-PCR products: Alu-PCR is an exceptional type of PCR reaction in which only a single primer is used, corresponding to a sequence which is chosen to be close to the end of *Alu* repeat consensus sequence (Nelson*et al.*, 1989). Since the *Alu* repeat is present at very high frequency in human DNA, two such repeats will often be found in close proximity and sometimes *in opposite orientations*. A single *Alu*-end primer can bind to each of two closely located, oppositely orientated *Alu* repeats and permit amplification of the sequence between them.

If the starting human DNA is a complex genomic clone such as a BAC or YAC clone, the *Alu*-PCR reaction will generate a diagnostic series of bands of specific sizes which can be compared with equivalent products from other library clones to check for overlapping inserts. In addition, since *Alu*-PCR products often consist of single-copy DNA (except for the incorporated primers), individual or pooled products from a single starting clone can be used as hybridization probes for screening other YACs or collections of inter *Alu*-PCR products derived from them. A similar type of amplification is possible for any class of highly repetitive interspersed repeat provided the repeat is suitably frequent so that neighboring elements are often found within a few kilobases, and there is sufficient sequence conservation to design a useful consensus primer.

Restriction site-based fingerprinting: Long-established clone fingerprinting methods have involved restriction nuclease treatment of clones and separation of radioisotope-labeled restriction fragments on gels. Modern methods of clone fingerprinting using restriction nucleases seek to achieve a high degree of automation using fluorescence labeling and detection systems (fluorescent fingerprinting). As an example of this approach, the method of Gregory *et al.* (1997) involves digesting clones with *Hin*dIII and *Sau*3AI, specific end-labeling of the *Hin*dIII ends with fluorescently labeled dideoxyATP, and size fractionation on a model 377 automated DNA sequencer.

Clone gridding and increasing automation have facilitated large-scale physical mapping efforts

Using current technologies, physical mapping of complex genomes is arduous and continues to be hampered by various practical difficulties:

- Chimeric DNA clones. Cloning artifacts can mean that cells contain two noncontiguous pieces of DNA from the desired genome. This may occur as a result of *co-ligation* of two different restriction fragments, internal deletion and *cotransformation*. Yeast cells that have been transformed by YACs and used to propagate have often been much more recombination proficient than typical *E. coli* hosts, and so YACs are often chimeric.
- Repetitive DNA. Arranging clones in a physical order means relying on some clone-specific features. DNA sequences that are repeated can make the task of ordering clones a difficult one, and may cause problems in sequencing DNA.
- Unclonable sequences. Some DNA sequences, including classes of repetitive DNA that are not well represented in bacterial and yeast cells, may not easily be propagated in such cells. A variety of different genomic libraries may need to be screened (including those constructed with vectors containing a low copy number origin of replication such as BAC and P1 libraries).

A major requirement for efficient physical mapping of complex genomes is the need for gridded DNA libraries. Following transformation of the host cells, individual bacterial or yeast colonies are picked into individual wells of microtiter dishes and stored in numerous multiwell dishes.

As a result, each cell clone can be assigned an identifying grid coordinate, comprising dish number, row number and column number. When libraries need to be screened by hybridization with a labeled DNA probe, it is again most efficient to have previously prepared high density gridded colony filters in which the colonies have been spotted on to the appropriate membrane in a gridded fashion with a uniform number of rows and columns. This simplifies identification of hybridization-positive colonies. More importantly, it facilitates integration of the results from numerous independent research groups.

This is possible because of increasing automation. Robots have been designed to pick and spot colonies, and numerous copies of reference libraries can be distributed, either as cell cultures or, more conveniently, as sets of colony filters and sent to researchers throughout the world. Hybridization data can then be sent back from individual laboratories to the distributing centers and the fixed geometry of the library coordinates simplifies the task of integrating the information received.

The Human Genome Project is one of several projects to benefit from such technologies, and sets of hybridization filters containing gridded colonies for total genome YAC libraries, chromosome-specific cosmid libraries and other libraries have been distributed extensively. In addition, pools of yeast colonies have been distributed in agarose plugs to permit PCR typing for a sequence of interest. Again, this is done in a hierarchical coordinate manner to simplify the screening of libraries.

Constructing transcript maps and identifying genes in cloned DNA

A wide variety of different methods can be used to identify genes in cloned DNA

A primary goal for physical mapping is identifying the locations of genes within a clone contig that has been localized to a specific chromosomal region. In principle, two major features permit the DNA of genes to be distinguished from DNA that does not have a coding function.

- Expression. All active genes are capable of making an RNA product, which, in the vast majority of cases, is mRNA. Mammalian genes usually contain exons and so the initial RNA transcript usually needs to undergo splicing.

Sequence conservation. Because genes execute important cellular functions, mutations which alter the sequence of the product will often be disadvantageous and are rapidly eliminated by natural selection. The sequence of coding DNA and important regulatory sequences is therefore more strongly conserved in evolution than that of noncoding DNA. In addition, premature termination codons in coding DNA are selected against. This means that genes often contain comparatively long open reading frames (ORFs); in noncoding DNA, the DNA triplets corresponding to termination codons are not selected against and ORFs are usually comparatively short. Some exons, however, may be quite small but can often be detected using computer programs to analyze the relevant DNA sequence.

In addition, vertebrate genes are often associated with CpG islands. These features have permitted a variety of different methods for identifying genes in cloned vertebrate DNA (Monaco, 1994) of which the most commonly used.

Genes can be identified easily by hybridizing DNA clones against northern blots, cDNA libraries, zoo blots and Southern blots of genomic DNA digested with rare-cutter restriction endonucleases. Of the various methods of identifying genes, the simpler methods are normally used as a first line of attack. In each case, the method involves using the genomic DNA clone as a hybridization probe. Inserts in cosmid clones and subclones are often useful hybridization probes. Larger clones such as YACs occasionally can be used as hybridization probes, but are generally less amenable to such approaches.

Hybridization to RNA/cDNA: A candidate DNA clone can be hybridized against a northern blot containing a panel of mRNA or total RNA samples isolated from a variety of different tissues (brain, heart, lung, liver, kidney, etc.). Positive hybridization may indicate the presence of a gene within the cloned fragment and may suggest a suitable cDNA library for screening. This approach has been facilitated enormously by using whole cosmid clones as hybridization probes in competition hybridization. They may fail, however, for two reasons. First, significant expression of the gene may be restricted to a cell population or developmental stage that is not represented in the northern blot panel or those cDNA libraries which are selected for screening. Additionally, there may be a problem with the intensity of the hybridization signal.

For example, the proportion of exon sequence in the probe may be very low (the average exon size is only about 200 bp and a 40 kb insert in a cosmid may by chance contain only one exon). If this is the case and if the gene is not strongly expressed in the relevant tissue (so that it is not well represented in the RNA samples or cDNA libraries which are being screened), detection of positive hybridization signals may be difficult. Occasional transcribed repeats may also be problematic.

Zoo blot hybridization: Coding DNA sequences are subject to considerable selection pressure to conserve biologically important sequences. By contrast, noncoding DNA sequences accumulate mutations comparatively rapidly and are not well conserved between species.

A zoo blot is a Southern blot of genomic DNA samples from a wide variety of different species. A genomic DNA clone which shows positive hybridization signals against the DNA of a variety of different species would be expected, therefore, to contain coding DNA sequences that have been strongly conserved during evolution. Some mammalian genes, often with a crucially important function in development, are so highly conserved that they will show significant hybridization signals with evolutionarily distant species such as yeast, *Drosophila* and *Caenorhabditis elegans*. Others may only show significant hybridization signals to mammals.

CpG island identification: CpG islands are short ~1 kb) hypomethylated GC-rich sequences which are often found at the 5' ends of vertebrate genes. In the human genome, an estimated 56% of genes are associated with such sequences. They include all examples of housekeeping genes and genes that are widely expressed, and a significant portion, perhaps about 40% of genes which show tissue-specific or restricted expression patterns.

The % (G + C) of CpG island sequences generally exceeds 60% and they have a high concentration of the CpG dinucleotide (typically the CpG frequncy is 10–20 times greater than that in the bulk vertebrate DNA). As a result, CpG islands often have restriction sites for a variety of rare-cutter restriction nucleases which cleave at GC-rich sequences containing one or two CpG dinucleotides. For example, each of the enzymes *Sac*II (CCGCGG), *Eag*I (CGGCCG) and *Bss*HII (GCGCGC) is expected to cut, on average, about 1.2 times within an island (see Cross and Bird, 1995), but very rarely outside islands. Close clustering of such restriction sites in genomic DNA is often indicative of a CpG island, and can be identified by using genomic DNA clones as hybridization probes against southern blots of suitably digested genomic DNA samples. Note, however, that the method is not applicable to the substantial number of genes which have no associated CpG islands.

Exon trapping identifies expressed sequences by using an artificial RNA splicing assay

The gene identification methods described above have their limitations: restricted expression patterns of some genes may make them difficult to identify; genomic DNA clones may give very weak hybridization signals if the percentage of exon sequence is very low; and many genes which are expressed in a tissue-specific or restricted manner do not have associated CpG islands. An alternative is to identify a gene by virtue of the ability of its exons to engage in an artificial RNA splicing assay. RNA splicing involves fusion of exonic sequences at the RNA level and excision of intronic sequences. Spliceosomes are able to accomplish this *in vivo* by recognizing certain sequences at exon/intron boundaries: a *splice donor* sequence at the junction between an exon and its downstream (3') intron, and a *splice acceptor* sequence at the junction between an exon and its upstream (5') intron. A cosmid or other suitable genomic DNA clone containing an internal exon flanked by intronic sequences will therefore contain functional splice donor and acceptor sequences.

Exons can be identified in cloned genomic DNA by subcloning the DNA into a suitable expression vector and transfecting into an appropriate eukaryotic cell line in which the insert DNA is transcribed into RNA and the RNA transcript undergoes RNA splicing. Such techniques are known as exon trapping (often called exon amplification if a PCR reaction is employed to recover the exons from a cDNA copy of the spliced RNA). For example, in the method of Church *et al.* (1994), the DNA is subcloned into a plasmid expression vector pSPL3 which contains an artificial minigene that can be expressed in a suitable host cell. The minigene consists of:

- a segment of the simian virus 40 (SV40) genome which contains an origin of replication plus a powerful promoter sequence;
- two splicing-competent exons separated by an intron which contains a multiple cloning site;
- an SV40 polyadenylation site.

The recombinant DNA is transfected into a strain of monkey cells, known as COS cells. COS cells were derived from monkey CV-1 cells by artificial manipulation, leading to integration of a segment of the SV40 genome containing a defective origin of SV40 replication. The integrated SV40 segment in COS cells allows any circular DNA which contains a functional SV40 origin of replication to replicate independently of the cellular DNA. Transcription from the SV40 promoter results in an RNA transcript which normally splices to include the two exons of the minigene. If the DNA cloned into the intervening intron contains a functional exon, however, the foreign exons can be spliced to the exons present in the vector's minigene. After making a cDNA copy using reverse transcriptase, PCR reactions using primers specific for vector exon sequences should distinguish between normal splicing and splicing involving exons in the insert DNA.

Genes in complex DNA clones, such as YACs, can be identified by forming heteroduplexes with cDNAs and by PCR-based methods of amplifying CpG island sequences

The techniques described in the previous sections have been popular when dealing with inserts from comparatively small genomic clones, such as cosmid clones or subclones. In some cases, however, attempts have been made to identify genes within large genomic clones, notably YAC clones. Two popular methods are cDNA selection and island-rescue PCR.

cDNA selection/capture: Direct cDNA selection techniques (also called direct selection and cDNA selection) involve forming genomic DNA/cDNA heteroduplexes by hybridizing a complex cloned DNA, such as the insert of a YAC, to a complex mixture of cDNAs, such as the inserts from all cDNA clones in a cDNA library (Lovett, 1994).

The principle underlying the technique is that cognate cDNAs corresponding to genes found within the YAC will bind preferentially to the YAC DNA; several rounds of hybridization should lead to a huge enrichment of the desired cDNA sequences, enabling the identification of the corresponding genes. Considerable blocking of repetitive DNA sequences is required. Early approaches used immobilized YACs, but more modern approaches have used a solution hybridization reaction and biotin- streptavidin capture methods. Like all expression-based systems, the method depends on appropriate levels of gene expression (the cognate cDNAs should not be too rare in the starting population).

Additionally, genes containing very short exons may be missed because the heteroduplexes formed with cognate cDNAs may not be sufficiently stable. Another problem is that cDNAs may bind to pseudogenes which show a high degree of homology to the cognate functional genes.

Island rescue PCR: Island rescue PCR (IRP) can permit selective amplification of CpG island sequences from human YACs. The method depends on the high copy number of *Alu* repeats (so that there is a high chance of an *Alu* repeat in the vicinity of a CpG island) and the frequent occurrence of restriction sites for rare-cutters such as *Bss*HII in CpG islands. The YAC DNA is cut with a suitable rare-cutter restriction nuclease and fragments are ligated to a vectorette linker with a suitably complementary overhang. CpG island-*Alu* PCR is then possible using an *Alu*-specific primer and a vectorette-specific primer.

BIBLIOGRAPHY

Adelberg, Edward A.: *Papers on Bacterial Genetics*, Boston, Brown and Company, 1960.

Ausubel, F.M.: *Current Protocols in Molecular Biology*, New York, John Wiley and Sons, 1989.

Bains, William: *Biotechnology: From A to Z*, New York, Oxford University Press, 1998.

Barnum, Susan R.: *Biotechnology: An Introduction*, Belmont, Thomson/Brooks/Cole, 2005.

Bauer MW: *Biotechnology-the Making of a Global Controversy*, Cambridge, Cambridge University Press, 2002.

Brandwein, P.F.: *Sourcebook for the Biological Sciences*, San Diego, Harcourt Brace JOvanovich, 1986.

Broach, J.R.: *The Molecular Biology of the Yeast*, Cold Spring Harbor, Cold Spring Harbor Laboratory, 1981.

Chrispeels, Maarten : *Plants, Genes and Crop Biotechnology*, Sudbury MA, Jones and Barlett Publishers, 2003.

Clark, J.M.: *Experimental Biochemistry*, New York, W.H. Freeman and Company, 1977.

Conko, Gregory: *The Frankenfood Myth: How Protest and Politics Threaten the Biotech Revolution*, New York, Praeger, 2004.

Cox, Earl D. : *Beyond Humanity: Cyber Evolution and Future Minds*, Roackland, Charles River Media, 1996.

David Sadava: *Plants, Genes and Crop Biotechnology*, Sudbury MA, Jones and Barlett Publishers, 2003.

DeGregori, Thomas R.: *Bountiful Harvest: Technology, Food Safety and the Environment*, Washington DC, Cato Institute, 2003.

Dixon, Dougal: *After Man-A Zoology of the Future*, New York, St. Martin's Press, 1981.

Dodds, John H.: *Plant Genetic Engineering*, New York, Cambridge University Press, 1985.

Fedoroff, Nina V.: *Mendel in the Kitchen: A Scientist's View of Genetically Modified Foods*, Washington, Joseph Henry Press, 2004.

Friedberg, E.C.: *DNA Repair*, New York, WH Freeman and Company, 1985.

Fumento, Michael: *Bioevolution: How Biotechnology is Changing Our World*, San Francisco, Encounter Books, 2003.

Goodman, D.C.: *From Farming to Biotechnology: A Theory of Agro-industrial Development*, Oxford, Blackwell, 1987.

Horn, Toby M.: *Working with DNA and Bacteria in Precollege Science Classrooms*, Reston, National Association of Biology Teachers, 1993.

Kurzweil, Ray: *The Age of Spiritual Machines*, New York, Penguin Books, 1999.

Miller, Henry I. : *The Frankenfood Myth: How Protest and Politics Threaten the Biotech Revolution*, New York, Praeger, 2004.

Moravec, Hans: *Mind Children: The Future of Robot and Human Intelligence*, Cambridge, Harvard University Press, 1988.

Morholt, E. : *Sourcebook for the Biological Sciences*, San Diego, Harcourt Brace JOvanovich, 1986.

Old, R.W. : *Principles of Gene Manipulation*, London, Blackwell Scientific Publications, 1989.

Postman, Neil: *Technopoly: The Surrender of Culture to Technology*, New York, Vintage Books, 1992.

Retzer, W.J.: *Biotechnology Workbook*, Englewood Cliffs, Prentice Hall, 1991.

Sayler, G.S., Fox, R., and Blackburn, J.W.: *Environmental Biotechnology for Waste Treatment*, Plenum Press, New York, 1991.

Slater, R.J.: *Experiments in Molecular Biology*, Clifton, Humana Press, 1986.

Switzer, R.L.: *Experimental Biochemistry*, New York, W.H. Freeman and Company, 1977.

Walden, Richard: *Genetic Transformation in Plants*, England, Open University Press, 1988.

INDEX

□□□